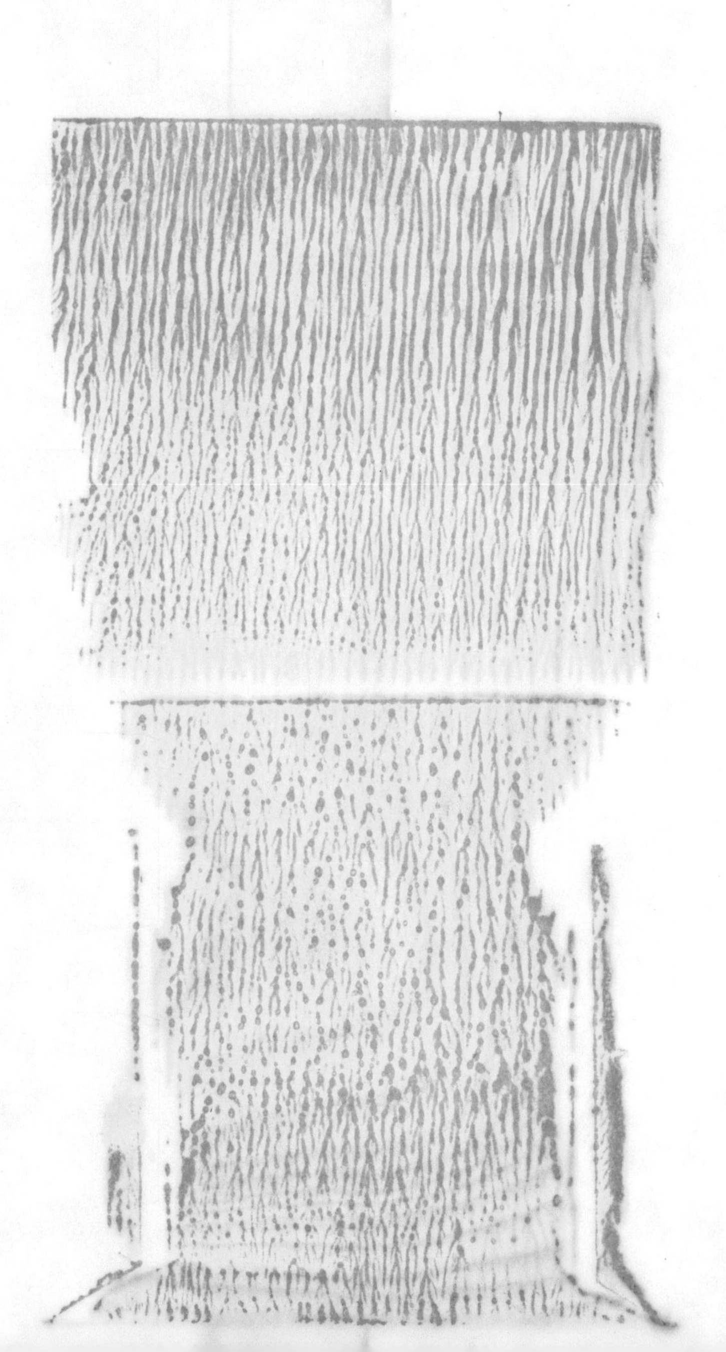

The Measure of the Universe

Some Recent Books on Science
by ISAAC ASIMOV

EXTRATERRESTRIAL CIVILIZATIONS *(1979)*

A CHOICE OF CATASTROPHES *(1979)*

IN THE BEGINNING *(1981)*

THE SUN SHINES BRIGHT *(1981)*

EXPLORING THE EARTH AND THE COSMOS *(1982)*

ASIMOV'S BIOGRAPHICAL ENCYCLOPEDIA OF SCIENCE AND TECHNOLOGY (2ND REVISED EDITION) *(1982)*

COUNTING THE EONS *(1983)*

THE MEASURE OF THE UNIVERSE *(1983)*

Illustrations by

Roger Jones

ISAAC ASIMOV

The Measure of the Universe

OUR FOREMOST
SCIENCE WRITER
LOOKS AT THE WORLD
LARGE AND SMALL

1817

HARPER & ROW, PUBLISHERS, New York
Cambridge, Philadelphia, San Francisco,
London, Mexico City, São Paulo, Sydney

 For information address Harper & Row, Publishers, Inc., 10 East 53rd Street, New York, N.Y. 10022. Published simultaneously in Canada by Fitzhenry & Whiteside Limited, Toronto.

FIRST EDITION

Designed by Ruth Bornschlegel

Library of Congress Cataloging in Publication Data

Asimov, Isaac, (date)
The measure of the universe.

Includes index.
1. Physical measurements—Popular works. I. Title.
QC39.A76 1983 530.8 82-48654
ISBN 0-06-015129-3

---

83 84 85 86 87 10 9 8 7 6 5 4 3 2 1

Dedicated to Angel and Satan,
my grand-cats

CONTENTS

He gave man speech, and speech created thought,
Which is the measure of the universe

Percy Bysshe Shelley,
Prometheus Unbound

INTRODUCTION

The Universe is so vast and the atom is so small that in neither case does it seem that we can grasp the dimensions involved.

We can use numbers to express the size in either direction, to be sure, for numbers can be used to represent any quantity, however large or however small. The trouble is that though numbers may be used to represent vastness or minuteness, they are then no more comprehensible than the things themselves. They can be manipulated and worked with far more easily than the objects themselves can—that is their advantage—but they cannot be pictured any more easily.

It seems a shame that we should find ourselves so unable to live with the Universe and its parts, especially since scientists have to work with them and are driven to ever greater extremes of measurement by curiosity and by the need to understand.

Very likely, nothing will solve the difficulty entirely. There is no way of grasping the size of the Universe, or of an atom, in the same way we can understand that of a breadbox, or a cat, or anything else we are accustomed to that has a size sufficiently similar to our own.

It occurs to me, though, that any measurement, however extreme, will seem less strange if we approach it a little at a time, by way of a series of regular steps. After all, we may not be able to jump to the top of a building in a bound, as Superman can, but even the least Supermanish of us can climb a ladder, or a flight of stairs, and arrive at the top, step by step.

What I wish to present in this book, then, is a number of different ladders, up which or down which we can climb, in order to get a somewhat better intuitive grasp of a few of the Universe's extremes.

The Ladder of Length
UPWARD

UPWARD

STEP 1

1 metre (10^0 m)

Suppose we begin by considering the Universe in the light of measurements of length. What is the length of a line that stretches from here to a star? What is the length of a line that stretches from one side of an atom to the other?

In order to express such lengths, we will have to use a "unit of measurement," some familiar length which we can then use in multiples and in subdivisions. For instance, any adult American would have a rough idea of what kind of length a "mile" represents, so that we can speak of lengths in terms of so many miles, or of such and such a fraction of a mile. Something might be 4 miles away, or 0.5 miles away or 2,769.4 miles, or 1/1600 of a mile.

Or we can uses inches, or feet, or yards, always remembering that there are 12 inches to 1 foot, 3 feet to 1 yard, and 1,760 yards to 1 mile. In principle, we can use any of a number of units of length that have been used to measure distance in the past, or which may still be used, even today, in specialized cases. Examples of such units are "fathoms," "cubits," "leagues," "ells," "hands," and so on.

There is no point, however, in using units that a great many people are unfamiliar with. What's the use of saying that one town is 14 leagues from another if one doesn't know how long a league might be? That limits us, here in the United States, to the use of miles for long distances and inches for short ones. We can say that a certain star is so many trillions of miles away, or that a particular atom is so many billionths of an inch across.

This book is not, however, being addressed to an American audience only. It will, I hope, be translated into many languages, and in that case I should use units of measurement that are familiar to the whole world.

As it happens, virtually the whole world, except for the United States, uses a particular system of measurement called "the metric system," a system first established in France in the 1790s. What's more, the rules for using the metric system have been formalized and made uniform through international agreement in the 1950s. The new rules are termed, in French, *Systeme International d'Unites,* which, it is not hard to see, is "International System of Units" in English. The new rules are usually spoken of as the "SI version."

Increasingly, scientists are using the SI version, and I will use it in this book. This will not be too unfair to the United States since we cannot hold out forever against world usage. American scientists have, for many years, been using the metric system exclusively, and in fact, little by little, the United States as a whole is accepting it. It will do American readers good to get used to the system, and I will make it as painless as possible by giving the common American equivalent wherever that would be useful.

The SI unit of measurement is the "metre," from a Latin word meaning "measure." In English, the term has been commonly spelled "meter," and I have used that spelling myself until very recently. However, the SI system rigidly sets standards for spelling, pronunciation, abbreviation, and so on, in order that the scientific use of measurements be a truly international language with no chance of misunderstanding across language barriers.

This goal is an unexceptional one, and I will try to follow the rules even when (in my heart) I would rather not. For instance, I cannot help but think that "meter" is the better spelling, but the convention is "metre" and I will suppress my rebellion over the matter.

It is pronounced "MEE-ter" (all units in the SI system are, in English, accented on the first syllable, though there may be secondary stresses elsewhere), and it is symbolized "m" so that one can write "1 metre" or "1 m" with equal validity. (The "m" is a symbol and not an abbreviation, so you don't use a period.)

The metre is not a difficult unit for Americans to grasp, since it is not very different from the familiar yard. A metre is equal to 1.094 yards, or to just about 1 1/10 yards. A yard is equal to 0.9144 metres, or to just about 9/10 metre. For rough approximations, you may even use the yard and the metre interchangeably.

Since a metre is also equal to 3.281 feet, and to 39.37 inches, it can be helpful, as a rule of thumb, to treat a metre as equal to 3 1/4 feet, or 40 inches.

The metre can be compared to natural phenomena; sound waves, for instance. Suppose you start at middle C on the piano and move two white notes upward to E, which is the sound we usually sing as "mi" when we intone the scale. The sound wave associated with that note is just about 1 metre long.

Sound waves consist of air (or some other substance) being alternately compressed and expanded by some sort of vibration. There are also "electromagnetic waves" produced by the oscillation of an electromagnetic field. Such waves, of the type used in broadcasting television signals, are in the neighborhood of a metre in length. Waves of this length and longer are commonly called "radio waves" because their first practical use was in the transmission of radio signals.

But what is a metre in terms of phenomena as familiar as our body?

An unusually tall person (6 1/2 feet tall in American units) is 2 metres tall. If such a person were to stretch out his arm sideways, shoulder high, then the distance from his nose to the tip of his outstretched fingers would be about 1 metre. If he were walking normally, his stride (that is, the swing of each foot from a position behind the other to a position in front) would be about 1 metre.

Nevertheless, this connection between the metre and the human body is coincidental. The length of the metre, as I shall later explain, was obtained from a natural length that had nothing to do with the human body.

STEP 2

3.16 metres ($10^{0.5}$ m)

The previous section, "Step 1," was headed *1 metre (10^0 m),* while this one is headed *3.16 metres ($10^{0.5}$ m).*

Why have I increased the measurement from 1 metre to 3.16 metres as I moved one step up the ladder, and what is the meaning of figures such as 10^0 and $10^{0.5}$?

Let's consider: Suppose I constructed the ladder by adding some constant figure to 1 metre, over and over. We might add 1 metre each time, going from 1 metre, to 2 metres, to 3 metres, to 4 metres, and so on. This is an "arithmetic progression."

The higher one goes in an arithmetic progression, the less significant the addition becomes. It would be useful to consider, separately, distances of 1, 2, 3, and 4 metres, since each would have its own points of interest. By the time we reached higher figures, however, what could we say about 76 metres that had not been said about 75 metres? The situation would be even worse when we talked about 872 metres and then went on to 873 metres. Besides, consider that before we are done, we should be talking of billions of billions of metres, and would never really have a chance to do so if we move upward by steps of 1 metre.

Even if we went by bigger steps in arithmetic progression—1 metre, 101 metres, 201 metres, 301 metres and so on—the interest would fade as the numbers grew larger, for the steady addition of 100-metre lengths would grow steadily less significant—and it would still take forever to reach the final stage.

We might really stretch and go up by stages of 1,000,000,000—dealing with 1 metre; 1,000,000,001 metres; 2,000,000,001 metres; and so on. It would *still* take us too long to get to the end, even then, and by moving up 1,000,000,000 metres in the first step, we would skip a great many levels of figures that would be of extreme interest.

In short, no arithmetic progression can possibly be useful as a means of constructing a ladder of the Universe. It would be too long, and it would concentrate too much attention on the very large numbers at the farther end of the ladder and too little attention on the small numbers at the nearer end of the ladder—the end nearer to ourselves.

The alternative is to *multiply* each number by some particular figure to get the next number. This would be a "geometric progression." If the number by which we multiply is 2, we would have 1, 2, 4, 8, 16, 32, 64, 128, 256, 512, and so on.

A geometric progression is the proper way of building a ladder of the Universe. Not only does a geometric progression get larger much more rapidly than an arithmetic progression does and therefore give us hope of reaching really large numbers in a reasonable time; it also takes small steps at the lower end of the scale and larger and larger steps at the upper end, something that exactly matches our interest in the matter.

But what number ought we to use as a multiplier in order to build a particularly useful geometric progression?

The accident of the form in which our number system exists makes 10 a particularly simple multiplier. Thus, if we start with 1 and multiply by 10 each time, we have the series: 1, 10, 100, 1 000, 10 000, 100 000, 1 000 000, and so on. (It is customary in the United States to divide the digits in large numbers into groups of three separated by commas. In many other countries, however, commas are used as what *we* call "decimal points." To avoid confusion, the SI system recommends that such groups of three digits be separated simply by a space, and this I will do from now on.)

A geometric series based on 10 as a multiplier has an elegant simplicity about it; it is therefore very commonly used, and scientists speak of "orders of magnitude" in this connection. Two objects differ by an order of magnitude in some measured property if the value of that property in one is 10 times that of the other. There are two orders of magnitude difference if the measure of the property of one is 10×10, or 100 times the other; three orders of magnitude difference if the measure of the property of one is $10 \times 10 \times 10$, or 1 000 times the other; and so on.

If, however, we consider the series 1, 10, 100, 1 000, 10 000, 100 000, 1 000 000, and so on, the numbers, as they grow larger, take up considerable room, and it becomes difficult to be sure of the number of zeroes at a glance. For that reason, mathematicians have worked out more compact systems for representing such numbers.

Instead of writing the series as we just have, we can write it as 1, 10, 10×10, $10 \times 10 \times 10$, $10 \times 10 \times 10 \times 10$, and so on. The increasing numbers of tens grows steadily more unwieldy, to be sure, and the whole is even clumsier and less readily read and understood than the original. We don't have to write down each ten though; we merely have to number them.

Thus, 10^1 is one 10, standing all by itself; 10^2 is the product of two 10s multiplied together; 10^3 is the product of three 10s multiplied together; and so on. The 10, in numbers expressed this way, is the "base," and the upper number is the "exponent." A number such as 10^3 is called an "exponential number."

The usefulness of such exponential numbers is easily seen:

$$10^1 = 10$$
$$10^2 = 10 \times 10 = 100$$
$$10^3 = 10 \times 10 \times 10 = 1\,000$$
$$10^4 = 10 \times 10 \times 10 \times 10 = 10\,000$$
$$10^5 = 10 \times 10 \times 10 \times 10 \times 10 = 100\,000$$

You can see the regularity in this, and, without actually writing down tens, be sure that:

$$10^6 = 1\,000\,000$$
$$10^7 = 10\,000\,000$$
$$10^8 = 100\,000\,000, \text{ and so on.}$$

You can see that the exponent, in an exponential figure of this type, is always equal to the number of zeroes in the same number written out in full. Thus, 10^{51} would be, in full, 1 000 000 000 000 000 000 000 000 000 000 000 000 000 000 000 000 000 000. Clearly 10^{51} is a much briefer and a much less confusing way of writing the number.

The expression 10^1 can be read "ten to the first power," 10^2 as "ten to the second power," 10^3 as "ten to the third power," 10^4 as "ten to the fourth power," and so on. Often, in rapid speech among people familiar with the system, the word "power" is omitted, and the talk is of "ten to the fourth," "ten to the fifth," and so on. They might even refer to "ten to the four," "ten to the five," and so on.

What's more, 10^2 and 10^3 are rarely referred to as "ten to the second power" and "ten to the third power," but usually as "ten squared" and "ten cubed" for reasons that go back to geometry and with which we don't have to concern ourselves.

As for 10^1, that is rarely treated as an exponential figure. Since 10^1 equals 10, the exponent is almost invariably omitted so that 10^1 is written simply as 10.

Exponential numbers are much more than merely a convenient brief way of writing large numbers. They also greatly simplify multiplication and division. Thus, 10 000 × 100 000 = 1 000 000 000, as you can see for yourself if you care to work out the multiplication in full in whatever way you choose. Put it into exponential figures and you have $10^4 \times 10^5 = 10^9$.

Notice that $4 + 5 = 9$. It would seem that, in the particular multiplication cited in the previous paragraph, we added the exponents in the two numbers being multiplied to get the exponent in the product. It turns out that this is a general rule for exponential numbers. Instead of multiplying ordinary numbers, you turn them into exponential numbers and add the exponents.

Division is multiplication in reverse. Thus, 100 000/1 000 = 100. In exponential numbers, this is $10^5/10^3 = 10^2$. As you know, $5 - 3 = 2$. The general rule for exponential numbers is that division involves the subtraction of exponents.

Well, then, consider the following division: 1 000/1 000 = 1. That is perfectly straightforward and unquestioned. Suppose, though, we put it into exponential numbers. It becomes $10^3/10^3 = 10^?$. By the rule of exponent subtractions, since $3 - 3 = 0$, $10^3/10^3$ should be equal to 10^0. Thus, the same problem in division gives us two answers: 1 and 10^0. The only way of keeping mathematics consistent is to suppose that these two answers are equal, and that $10^0 = 1$.

In ordinary affairs, no one ever uses 10^0 in place of 1, but mathematicians sometimes do so when the exponential number preserves symmetry, or allows an arithmetical rule to be made general. I am using 10^0 in these ladders of the Universe for the sake of symmetry.

Thus, Step 1 above was listed as "1 metre" and then, in parentheses, as "10^0 m," saying the same thing in exponential number and symbol.

But what about Step 2? Why did I not jump an order of magnitude to "10 metres (10^1 m)"?

Multiplying by 10s, and moving up an order of magnitude at a time, takes steps that are too large for my purpose, at least in this particular ladder.

I could multiply by 5s instead, but that would give me a series that is clumsy: 1, 5, 25, 125, 625, 3 125, 15 625, and so on. To avoid that, I could use a hybrid of fives and tens, thus: 1, 5, 10, 50, 100, 500, 1 000, 5 000, and so on.

This, however, leaves us with steps of unequal size. Beginning with 1, we are multiplying first by 5, then by 2, then by 5, then by 2, and so on.

What we really want is a way of taking two multiplications to reach 10, but having each multiplication use the same multiplying value. We would then be advancing each time by half an order of magnitude.

Thus, I would multiply 1 by some number, a, which would give me a, of course (any number multiplied by 1 has a product that is the original number, unchanged). Then I would multiply a again by a and have 10. I am therefore looking for a solution to the equation $a \times a = 10$. Whenever a small number multiplied by itself gives a larger number, the small number is said to be the "square root" of the larger number. What we are looking for is the square root of 10.

Mathematicians know how to calculate square roots. Usually, square roots are "irrational"; that is, they are neither whole numbers nor fractions, but can only be expressed as an unending decimal. In such a decimal there is no pattern, so that you can never predict what the next digit will be unless you figure it out. Still, if you want to take enough trouble, you can calculate any square root to as many decimal places as you wish.

The square root of 10 is equal to 3.1622776 . . . and so on. We don't have to worry about the additional places under ordinary conditions since $3.1622776 \times 3.1622776 = 9.9999996$, which is certainly close enough to 10 for our purposes. In fact, 3.16 is close enough, since $3.16 \times 3.16 = 9.9856$.

Thus, we have a series: 1, 3.16, 10, 31.6, 100, 316, 1 000, 3 160, and so on. In this series, all the steps are equal since each number is multiplied by 3.16 in order to get the next higher number. Since, in this series, there are two steps in going from 1 to 10, and from 10 to

100, and from 100 to 1 000, and so on, we are indeed moving up the ladder in steps of half an order of magnitude.

What about exponentials? Suppose we want to put the equation $3.16 \times 3.16 = 10$ (approximately) into exponential numbers. We don't know the exponential form of 3.16, so let's just call it 10^a. We therefore have the equation $10^a \times 10^a = 10^1$. By the rule of exponents, we know that in such an equation $a + a = 1$. In that case a must equal 1/2 or 0.5. Therefore, the square root of 10, which is approximately 3.16, can be expressed as $10^{0.5}$. That is why Step 2 in the ladder of length is headed "3.16 metres ($10^{0.5}$ m)."

The series 1, 3.16, 10, 31.6, 100, 316, 1 000, 3 160 . . . can be expressed exponentially as 10^0, $10^{0.5}$, 10^1, $10^{1.5}$, 10^2, $10^{2.5}$, 10^3, $10^{3.5}$ . . . and so on.

Having explained all that (which supplies us with almost all we need to know about exponentials and orders of magnitude for the rest of the book), let's consider the significance of a measurement of 3.16 metres.

This is just about equal to 10 3/8 feet, so that a Step-2 length is about the height of a story in a modern building and outranges the tallest normal human being, if we ignore fanciful legends of giants in myth and legend. The tallest human being of whom there is record was Robert P. Wadlow. He was not, alas, normal, but was afflicted with "giantism," a hormonal disorder. He never stopped growing, and in 1940, when he died at the age of 22, he was about 2.75 metres (9 feet) tall. Even this is well short of the Step-2 length.

There is such a thing as a near-human giant in reality, too. The largest ape that ever lived is now extinct but is known from fossil teeth, jawbones and other scraps. It is *Gigantopithecus* (Greek for "giant ape"). It looked very much like an outsize gorilla, and when it stood erect it was perhaps 2.75 metres tall—just the height of Wadlow, but in the case of *Gigantopithecus,* it was a normal height and not a disorder.

There are animals that are taller than any primate, living or extinct, and that reach the Step-2 level or beyond. The height of a male African elephant, at the shoulders, was as high as 3.8 metres in one recorded case.

Where sound waves are concerned, one that is 3.16 metres long is a G, two octaves below the G of the ordinary scale. It is a majestic bass note (the longer the sound wave, the deeper the pitch).

An electromagnetic wave that is 3.16 metres long is still in the TV range, but is near the upper wavelength limit of that range.

In the incredible wilderness of measurements used by humanity, I will occasionally mention one or two in the range under discussion that are still occasionally used in the United States. A "rod" (or a "pole," or "perch"—all three taken from the long stick used as a standard for that length) is defined as 5 1/2 yards, which comes to 5.029 metres: somewhat beyond the Step-2 length, as you can see, but well below the Step-3 length we are about to come to.

STEP 3

10 metres (10^1 m)

When the metric system was first devised, the units received different names for each order of magnitude, the differences being indicated by prefix.

Thus, 10 metres was set equal to "1 dekametre" (pronounced, in English, DEK-uh-mee-ter). The prefix is symbolized as "da," to distinguish it from another prefix we'll come to later, which uses the simple "d" as a symbol. In that way, 1 dekametre can be written as 1 dam.

The prefix "deka" is the Greek word for ten, so that if one is acquainted with Greek, the meaning of "dekametre" is clear at once. Otherwise, one must simply memorize the prefix and the meaning. Once that is done, however, it works for any kind of metric measure. A dekapoise is 10 poises, a dekawatt is 10 watts, and so on.

Actually, the dekametre, or any metric measure using the "deka" prefix, is rarely used. In the SI system, prefixes are used only for every *three* orders of magnitude, since the total spread of measures is far greater than that envisaged by the originators of the metric system. If a separate prefix were used for *every* order of magnitude through the spread we now use, the matter would become unnecessarily and impractically complicated. However, having mentioned the "deka" prefix (as I will others of the sort), I will use it henceforth, though perhaps not invariably.

A length of 10 metres corresponds approximately to the height of a three-story building. There is no living animal that is that tall. The giraffe is the tallest living animal, and the tallest giraffe ever measured had the tip of its horns only 5.8 metres above the ground when it was standing erect, which makes it half again as tall as an elephant. (To put that into human terms, the highest pole vault achieved by a human being is about 5.7 metres.)

If we include extinct animals, the very largest dinosaurs, such as the brachiosaur, could raise their long necks until the top of the head was nearly 12 metres above the ground.

Measurements need not be made in the vertical direction only.

The largest elephants may be nearly 10 metres long, if the trunk and tail are both extended. There have been reports of snakes or alligators approaching the 10-metre mark, but those are probably exaggerated. It is likely that 10 metres in any direction is about the longest any living land animal is likely to measure.

A sound wave 10 metres long is close to the deepest sound that can be heard by human ears. Sound waves with still longer waves are too deep in pitch to affect the mechanism of the human ear. They are "infrasonic" or "subsonic" waves, and we will follow them no longer.

Electromagnetic waves of this length are in the "shortwave" radio region. They are so-called because the waves are shorter than those used in ordinary radio. Waves in the shortwave region are also used in "frequency modulation" (FM) radio.

STEP 4

31.6 metres ($10^{1.5}$ m)

By moving up another step we have left the world of land animals far behind. A length of 31.6 metres is the height of a 10-story building, and even the tallest brachiosaur can reach only 4 stories high.

In the sea, however, are animals that are not affected by gravity, since seawater buoys them to near-weightlessness whatever their size. They do not have to face the problems of supporting their great weight on legs of ever-increasing massiveness. (Elephants must have their bodies resting on virtual tree trunks, straight up and down for extra strength, and although they can move with surprising speed, they cannot jump—not an inch. The largest dinosaurs have their legs even thicker and more massive in proportion, and it may be that flesh and bone simply cannot support anything larger.)

Whales can be far larger than elephants, and can even outdo dinosaurs, without the necessity of legs at all. The water supports them. (If, however, a whale is beached, its own weight so compresses its lungs that it is unable to breathe, and suffocates.)

Even whales of moderate size exceed the largest of the land animals. The greatest of the toothed whales, the sperm whale, is just about 20 metres long, twice as long as the longest elephant, though still well below the Step-4 mark. (The sperm whale is the only whale with a throat wide enough to swallow a man whole.)

There are, however, whalebone whales, which do not live on large prey as the sperm whale does. (The sperm whale's favorite food is the giant squid, for which it is ready to dive to enormous depths and to remain underwater for over an hour.)

Whalebone whales live on tiny shrimp and fish which they strain out of the water by means of the cartilaginous fringes of "whalebone" that extend down from the roof of their mouths. Since small organisms are far more numerous than large ones, whales that feed on smaller organisms can prosper and grow to larger size than those that feed on larger ones.

The largest of the whalebone whales is the "blue whale" or "sulfur-bottom whale" of Antarctic waters. The largest blue whale ever captured and measured was 33.3 metres long.

As far as pure length is concerned, there have been occasional reports of tapeworms from whale intestines, and of jellyfish, that were longer. One jellyfish was reported to have tentacles that were 36 metres long. If some tentacles were stretched out in one direction and others in the opposite direction, an extreme length of some 70 metres might be recorded.

Yet jellyfish tentacles, and tapeworms, too, are very insubstantial, almost one-dimensional in nature. It is reasonably fair to say that the blue whale is the longest and largest animal of substance that lives on Earth, or that has *ever* lived on Earth.

It seems to me that for this reason alone the blue whale ought to be cherished by human beings, who must surely appreciate it as an awesome example of the potentiality of life—and yet its numbers are decreasing rapidly, and tragically.

To bring in the human world, a trained athlete can jump about 8.9 metres at most, or roughly the length of an elephant, and he can achieve a shotput of as much as 21.8 metres, or roughly the length of a sperm whale.

STEP 5

100 metres (10^2 m)

When the metric system was first established, 100 metres was set equal to a "hectometre," which is symbolized as "hm" and is pronounced HEK-toh-mee-ter. The prefix "hecto-" is from the Greek word "hekaton," meaning "hundred." The prefix "hecto-" is rarely used.

A length of 100 metres leaves the animal world behind. It is equal to the height of a 33-story building. My own apartment in Manhattan, on the thirty-third floor, is just about 100 metres above the ground.

Manhattan (since we've mentioned it) is for the most part divided from north to south by streets running due east and west. The distance between these streets (one city block) is just about 80 metres. A distance of 100 metres is therefore about 1 1/4 city blocks.

The world of life is not to be entirely ignored at the Step-5 range, however. Animals may be out of consideration, but there remains the world of plants. Taller than the longest whale are many species of trees. The tallest of all trees are the redwood trees that grow along the California coast. The tallest of these trees is reported to have attained a height of 112 metres, so that it is over four times as tall as the longest blue whale would be if it were standing on its tail.

There have been reports of still taller trees, either Douglas firs or Australian eucalyptus trees, with some estimates of heights of 150 metres or so, but these have never been verified. Of course, if the root-system of a tree is included, the total height from the topmost branch to the lowermost root of a very tall tree might well be as much as 200 metres.

Human achievements are still in competition at the Step-5 range of length. At throwing the javelin, in sports competitions, a record cast of 94 metres was achieved, which is roughly the height of a tall redwood.

Human beings can, in this respect, outdo any form of life in one way, since they can build structures with measurements longer than those that can be achieved by any other form of life, and have done it surprisingly early in the history of civilization, too.

The Great Pyramid of Egypt, built 4 500 years ago, is 146 metres high. It is an impressive pile of rocks that would be difficult to build today, yet it was built by sheer muscle power, making use of simple, but ingenious, devices in the hands of many thousands of human beings, who had to labor for many years at the task. (It was to remain the tallest human-made structure for three thousand years.)

Electromagnetic waves in the 100-metre range are at the upper limit of the shortwave radio band.

STEP 6

316 metres ($10^{2.5}$ m)

We have now reached a length that is four city blocks long, or the length of one of the long crosstown blocks in Manhattan, say from Fifth Avenue to the Avenue of the Americas. (I apologize for being so provincial as to make use of Manhattan frequently, but, first, I live in Manhattan and love it; second, it has a simple geometry; and third, of all cities it is perhaps the best known to nonresidents.)

The Step-6 range is still within the world of buildings. In the late

Middle Ages, the great cathedrals of western Europe began to exceed the Great Pyramid in height. The central tower of Lincoln Cathedral in England attained a height of 160 metres.

Of course, both the pyramids and the cathedrals were long labors inspired by religious enthusiasm. The first secular structure to surpass the cathedrals in height was the Washington Monument, an obelisk that is 170 metres tall. It was completed in 1884. (Here is one case where the use of the metric system obscures an interesting fact. The Washington Monument was deliberately built to a height of 555 feet and 5 inches.)

The Washington Monument was far outdone in 1889 with the completion of the Eiffel Tower, intended to commemorate the centennial of the French Revolution. It is a narrowing tower of steel, a skeleton of a building rather than a building, and was originally intended to be a temporary structure. It aroused a storm of protest from Parisians who felt it to be out of tune with the city and surpassingly ugly in its own right. It still stands, however, and Parisians, I imagine, have long since resigned themselves to it.

The Eiffel Tower is almost exactly 300 metres tall, twice the height of the Great Pyramid; and, for the first time, a man-made structure approached the Step-6 level of length.

None of these structures I mentioned, however, are intended for human beings to live and work in. They are tombs, memorials, places of worship.

Until the middle of the nineteenth century, dwelling and office buildings were almost never higher than four or five stories because the labor of climbing stairs came to be excessive after that. The first practical elevator came into use in 1859 and such devices were rapidly improved. With that, and with the fact that metal skeletons of iron and steel, rather than solid masonry, were used to bear the weight, tall buildings became truly practical.

Buildings so tall that they were hyperbolically called "skyscrapers" were built in several American cities but particularly on the island of Manhattan, which, for its solid phalanxes of tall buildings, came to be one of the unique places in the world.

By the beginning of the twentieth century, the skyscrapers were crowding the Step-6 level. The Metropolitan Life Insurance Building, completed in 1909, had its tower 213.34 metres (just 700 feet) above the ground.

The Woolworth Building, completed in 1913, had a height of 241 metres and remained the tallest office building in the world for a quarter of a century. (When I was a boy, "to jump over the Woolworth Building" was a proverbial expression for an impossible feat, and I used it frequently myself, though I hadn't the slightest idea what the Woolworth Building was.)

It was not till 1930 that an office building was constructed in Manhattan that outdid the Eiffel Tower and was, therefore, not only the tallest office building in the world, but the tallest human-made structure of any kind. It was the Chrysler Building, which had a height of 319 metres and thus was the first structure to rise above the Step-6 level of length.

Within a matter of months, however, it lost its lead to the Empire State Building, which was 381 metres high.

The Empire State Building held the record for a quarter of a century in its turn, until 1973, when the twin towers of the World Trade Center in Manhattan rose to a height of 411 metres. Then, the very next year, the lead passed to the Sears Tower in Chicago, which was 443 metres high. This remains now the tallest office building in the world, though there are television masts here and there which are taller, some of them attaining heights of well over 600 metres.

We can express the heights of these tall buildings in another and rather dramatic way. The Empire State Building is 4 3/4 city blocks tall. The Sears Tower is 5 1/2 city blocks tall.

The longest ocean liners are just about as long as skyscrapers are tall. At the time the Woolworth Building was completed, the British liner, *Lusitania,* was as long as that building was tall. The French liner, *France,* launched in 1961 and, so far, the longest ship built, was 315.5 metres long, just about as long as the Chrysler Building is high.

Electromagnetic waves that are 316 metres long are just about midway in the familiar "amplitude modulation" (AM) radio broadcasting band.

One fairly familiar common measure that approaches the Step-6 level is the "furlong." This was originally the length of a plowed furrow in a field and was set equal to 40 rods, or 220 yards. This is equivalent to 201.17 metres. This means that the Step-6 length of 316 metres is equal to 1.57 furlongs. (Tracks used in horse racing are still routinely measured in furlongs.)

STEP 7

1 000 metres (10^3 m)
1 kilometre (10^0 km)

When the metric system was first established, the prefix "kilo-" was established to signify 1 000 of a basic unit. This is from the Greek "chilioi" meaning "thousand." Thus, a "kilometre," symbolized "km" and pronounced "KIL-oh-mee-ter" (and *not* with the accent on the second syllable) is equal to 1 000 metres.

The kilometre is three orders of magnitude greater than the metre, and it is the rule that distinctive prefixes are used every three orders of magnitude, in order to avoid having to deal with figures that are too large to be convenient.

In this book, I will continue to use the basic unit—in this case the "metre"—in the heading of each step, for the sake of continuity. I will follow it, however, with the prefixed unit, and I will use only the prefixed unit in the body of the step.

Thus, electromagnetic waves that are 1 kilometre long (and I will not say 1 000 metres) are in the "longwave" region of the radio wave band. These are longer than those used in ordinary radio transmission. Longwaves have various specialized uses, and they can be indefinitely long. We will leave them behind and refer to radio waves no more after this step.

The kilometre is the preferred unit for longer distances in common life, as the metre is for shorter distances. Where the metre is the rough equivalent of the yard in common measure, the kilometre is the rough equivalent of the mile. The comparison is not as close as in the metre/yard case, however, for 1 kilometer = 0.6214 miles, and 1 mile = 1.609 kilometres.

More briefly, a kilometre is 5/8 of a mile, and a mile is 1 3/5 kilometres. To put it another way, 8 kilometres = 5 miles; to put it still another way, a kilometre is 3 281 feet or 1 094 yards, whereas a mile is 5 280 feet or 1 760 yards; to put it yet another way, a kilometre is 12.5 city blocks, whereas a mile is 20 city blocks.

A kilometre in New York City would represent the distance from Fifth Avenue to halfway between Eighth and Ninth avenues going westward; or from Fifth Avenue to First Avenue going eastward. Central Park, which stretches from Fifth Avenue to Eighth Avenue, is 0.8 kilometres wide.

By the time we've reached the Step-7 level, we've outranged anything human-made in height. Nothing human beings have ever constructed stretches a kilometre from bottom to top. On the other hand, there are natural elevations that rise far higher. A mountain that reaches a height of only 1 kilometre above sea level is not in the least impressive.

Ben Nevis, in west central Scotland, is the highest peak in Great Britain (certainly not a mountainous island), and it is 1.34 kilometres high.

At the Step-7 level, we can begin to talk about astronomical bodies.

They come in all sizes, from tiny motes of dust upward. If we restrict ourselves to bodies sufficiently impressive to receive names, then we can begin with small asteroids that, on occasion, skim close to Earth (as astronomical distances go) and that are therefore known, with considerable exaggeration, as "Earth-grazers." Of these, the best known to the general public is Icarus. That particular asteroid is well known, not only because of its periodic close approach to Earth, but because of its very close approach to the Sun at one end of its orbit.

The distance across Icarus, its "diameter," is, as near as can be estimated, a trifle more than 1 kilometre. That means if it were gently brought to rest in Manhattan, it would fill the lower third of Central Park (lapping over to my apartment house, probably) and would give the island a mountain about the size of Ben Nevis.

STEP 8

3 160 metres ($10^{3.5}$ m)
3.16 kilometres ($10^{0.5}$ km)

In the Step-8 range we exceed the mile, for 3.16 kilometres is about 2 1/4 miles. This represents the east-west width of the island of Manhattan, at its widest; and it is just short of the length of Central Park, which, extending as it does from 59th Street to 110th Street, is almost exactly 4 kilometres long.

There are some nonmetric measures of length, familiar to some Americans, that are in this range. There is in the first place the "nautical mile" (also called "geographic mile" and "sea mile") that is much used at sea. It is set by international agreement to a length of 1.852 kilometres, which makes it equal to 1.1516 ordinary miles (also called "statute miles") or 2 025.4 yards.

Why the difference between the nautical mile and the statute mile? The nautical mile is defined as 1 minute of arc of a meridian of longitude, and that comes out to a little more than a statute mile. It is convenient for navigators to have their miles fit in with the "circular measure" of degrees, minutes, and seconds used in traveling about a spherical Earth, and it matters less to them that it doesn't quite fit the mile that landlubbers use.

Then there is the "league," a unit of measure that varies in length from place to place. Usually, it is accepted as equal to 3 miles, which would also make it equal to 4.81 kilometres.

There are many mountains with heights that are in the Step-8 range. The highest mountain in Australia (the least mountainous of the continents) is Mount Kosciusko in southeastern New South Wales. It is only 2.23 kilometres high. Mount Cook, the highest mountain in New Zealand, on the west-central shore of South Island, is 3.76 kilometres high.

The Earth-grazer Geographos is about 3 kilometres in diameter and must resemble something like Mount Cook flying through space.

STEP 9

10 000 metres (10^4 m)
10 kilometres (10^1 km)

When the metric system was first devised, no attempt was made to set up prefixes for anything higher than 1 000 times the base unit. It is possible that it was felt that nothing higher would be needed. If so, this represented quite a shortsighted attitude. Modern scientists have found it necessary to invent prefixes for additional multiples far beyond the "kilo-."

These new prefixes occur every three orders of magnitude. Nothing in between is dealt with.

There is one tenuous exception. The Greek word "myrioi" meant "ten thousand," though often it was used more vaguely to represent an innumerable quantity. In the English language, we have "myriad," which usually applies to an innumerable quantity but is sometimes used to represent ten thousand.

The prefix "myria-" is therefore sometimes used to represent ten thousand times the basic unit. A "myriametre" (pronounced MIH-ree-uh-mee-ter) would be 10 000 metres or 10 kilometres. The myriametre has been rarely used in this manner, however, and the prefix is *not* an official part of the SI version. For that reason, I will not use it henceforward.

At the Step-9 level, we outrun the mountain heights.

The highest mountain in Europe is Mount Elbruz in the Caucasus mountain range, between the Black Sea and the Caspian Sea. It is only 5.64 kilometres high.

Mount Kilimanjaro in Tanzania is the highest mountain in Africa and is 5.89 kilometres high. Mount McKinley, located in south-central Alaska, is the highest mountain in North America. It does a little better and is 6.19 kilometres high. Mount Aconcagua, the highest mountain in South America, on the east-central border of Chile, does better still and is 6.96 kilometres high.

Aconcagua is the highest mountain in the world outside Asia, but in Asia there are at least 66 mountain peaks that are higher than Aconcagua. Of these the highest of all, and therefore the highest in the world, is Mount Everest, on the border of Nepal and Tibet. It is 8.848 kilometres above sea level and therefore still falls well short of the Step-9 level.

It may be unfair, however, to measure a mountain's height as the distance it reaches above sea level. Some mountains rise out of the sea, and it might make sense to measure the height their topmost peaks reach above the solid subsea surface on which they stand. The volcanic mountain, Mauna Loa, which makes up much of the island of Hawaii, has its topmost peak only 4.169 kilometres above sea level, but if its underwater slopes are included, its total height from base to peak is just about 10 kilometres.

Suppose that we now go in the other direction and inquire not about heights but about depths. How deep is the solid surface that underlies the sea?

The deepest part of the ocean is in the Marianas Trench, southeast of Guam in the Pacific. There, a depth of 11.033 kilometres has been measured, and this just exceeds the Step-9 level.

In fact, if we consider the distance from the deepest part of the ocean to the peak of the highest mountain, we see that the difference in level of Earth's solid surface reaches an extreme of 19.88 kilometres.

The variations of the Earth's surface in the vertical direction have been about played out at this step; but not so horizontal distances along the Earth's surface.

Thus, 10 kilometres is just about half the length of Manhattan Island; a distance from the tip of the Battery at the south, up to about 95th Street, near the northern end of Central Park.

If we move on to astronomical bodies, there are, at the Step-9 level, the diameters not only of small asteroids but also of small satellites —bodies that seem asteroidal in nature but circle particular planets rather than follow an independent orbit about the Sun. These may originally have been asteroids that were, at some time in the past, captured by the gravitational pull of the planets.

There is, for instance, Deimos, the smaller of the two satellites of Mars. It is not a spherical body (small objects of this size don't have a large enough gravitational field to force their own material into the compactness of a sphere). This means that the length of Deimos's diameter is not the same in all directions, as it would be if it were a sphere. Deimos's shortest diameter is 10 kilometres long and its longest is 16 kilometres.

STEP 10

31 600 metres ($10^{4.5}$ m)
31.6 kilometres ($10^{1.5}$ km)

Here at Step 10, we must bid Manhattan adieu, for its extreme length, north and south, is not quite 22 kilometres. The outer boroughs of Brooklyn and Queens (each of which is larger than Manhattan), taken together, have a maximum width of just about 30 kilometres.

If we imagined ourselves moving straight upward through the atmosphere, a distance of 31.6 kilometres would take us up out of the "troposphere," where the clouds and weather exist. We would be in the quiet stratosphere with more than 90 percent of Earth's air beneath us.

Phobos, the larger satellite of Mars, has its longest diameter equal to 28 kilometres, while its shortest diameter is 20 kilometres. If it were carefully placed upon New York, it would neatly cover Brooklyn and Queens. (Deimos, the smaller satellite, would cover Queens alone.)

The Earth-grazer Eros was the first of its class to have been discovered, back in 1898. It is a brick-shaped object, with its longest diameter about 36 kilometres and its shortest 16 kilometres. It is very like Phobos in size, but is a little bit longer and narrower.

STEP 11

100 000 metres (10^5 m)
100 kilometres (10^2 km)

At the Step-11 level, we must finally leave New York City behind, for its maximum width from the southwestern tip of Staten Island to the northern boundary of the Bronx is only some 47 kilometres. Los Angeles does somewhat better, for its maximum north-south length is 77 kilometres, but it is clear we are now beyond cities in general.

Indeed, we have left behind some of the smaller states and nations. The maximum length of the state of Rhode Island is only about 76 kilometres, and the European nation of Luxembourg does no better. However, the states of Delaware and Connecticut have maximum lengths of 170 kilometres each.

The planet Jupiter has four large satellites, and beyond them are at least eight or nine smaller bodies which are probably captured asteroids. The largest of these "outer satellites," and the first discovered, is now known as "Himalia." It is about 100 kilometres in diameter, right at the Step-11 level, and if placed on Earth's surface would cover two-thirds of Connecticut.

STEP 12

316 000 metres ($10^{5.5}$ m)
316 kilometres ($10^{2.5}$ km)

The northern boundary of Maryland, bordering on Pennsylvania, is known as the Mason–Dixon line because it was surveyed by Charles Mason and Jeremiah Dixon between 1763 and 1767. It later achieved a sad fame in American history as symbolizing the entire border between the free states and the slave states, a symbol it took the bloody Civil War to remove. The actual Mason–Dixon line is 312 kilometres long, and it is almost exactly a Step-12 length.

A bit longer is the northern boundary of California, bordering on Oregon. That east–west line is 355 kilometres long.

Here are some other lengths in the Step-12 range. From New Orleans to the northern border of Louisiana is 360 kilometres in a direct line. From Little Rock, Arkansas, to Jackson, Mississippi, is 336 kilometres.

In England, the distance from Plymouth northward to York in a direct line is 360 kilometres. In France, the distance from Paris eastward to Strasbourg is 400 kilometres. In the Soviet Union, the distance from Moscow westward to Smolensk (near where I was born) is 400 kilometres. From Tokyo to Osaka is also 400 kilometres.

The Step-12 range also brings us to the larger asteroids. Hygeia, the tenth asteroid to be discovered, is about 450 kilometres in diameter. Only three asteroids are larger than this.

STEP 13

1 000 000 metres (10^{6} m)
1 megametre (10^{0} Mm)

When the metric system was first devised two centuries ago, no provision (as I said before) was made for a prefix larger than "kilo-." Nowadays, however, a number of prefixes have been invented, made official, and been accepted, at intervals of three orders of magnitude. Thus, if we consider 1 000 000 metres (1 000 kilometres), it is clearly time for a new prefix.

The prefix used is "mega-," pronounced "MEG-uh" and symbolized "M." The capital letter is used because the small "m," as we shall see later, had been preempted for another prefix that has been in use much longer. The prefix "mega-" comes from the Greek word "megas," meaning "great." The prefix always signifies 1 000 000 (one million) times a basic unit, so 1 megametre = 1 000 000 metres = 1 000 kilometres.

The distance from Chicago to Washington, D. C., in a direct line (following the curvature of the Earth's surface, of course) is 960 kilometres, or 0.96 megametres. From Paris to Vienna, the distance is 1.04 megametres, and from Moscow to Stockholm, it is 1.15 megametres.

The Step-13 distance takes us to the limit of the asteroids. The largest of these bodies, and the first discovered, is Ceres. Its diameter, at most recent reports, is 1.003 megametres.

We have also reached the diameters of middle-sized satellites. The fifth largest of Saturn's satellites is Tethys; it has a diameter of 1.05 megametres. The planet beyond Saturn is Uranus, which has five known satellites of which the largest is Titania. The diameter of Titania is 1.04 megametres.

STEP 14

3 160 000 metres ($10^{6.5}$ m)
3.16 megametres ($10^{0.5}$ Mm)

There are very few straight-line boundaries on the political map. Most boundaries have been set by natural barriers combined with the complexities of thousands of years of war and diplomacy, so a straight line would be too simple. The new nations of the United States and Canada, occupying relatively empty lands, could manage to be simple.

The longest straight-line international line in the world is part of the boundary between the United States and Canada. (It is not actually a straight line on the globe, for it does not follow the curve of a "great circle," but it is a straight line, due east and west, on the Mercator projection, which is the usual form in which we see a world map.) That boundary, running from Puget Sound in Washington State to Lake of the Woods in Minnesota, is about 3.2 megametres long.

Only one straight line on the map does better. It is a provincial boundary within Canada. From the Alaska Panhandle to Hudson Bay, the northern boundary of British Columbia, Alberta, Saskatchewan, and Manitoba falls on a straight line (at least on the Mercator projection) running due east and west for about 4.0 megametres.

Here are some distances along the Earth's surface that are in the Step-14 region. These distances are taken along great circle routes.*

New Delhi to Bangkok	2.92 megametres
Berlin to Cairo	2.89 megametres
Washington, D.C., to Caracas	3.29 megametres
Tokyo to Hong Kong	2.88 megametres
San Francisco to Honolulu	3.86 megametres
Moscow to Madrid	3.45 megametres
Montreal to Mexico City	3.73 megametres
Rome to Teheran	3.41 megametres.

At the Step-14 level, we are moving into the realm of the larger satellites. Of Jupiter's four large satellites, the smallest is Callisto, which has a diameter of 3.13 megametres.

Our own Moon is somewhat larger, with a diameter of 3.48 megametres (about the distance from Moscow to Madrid).

Jupiter's satellite, Io, the nearest of the large satellites to Jupiter, has recently made news headlines because our rocket probes have discovered active volcanoes upon it. (It is the only world other than Earth itself known to have them.) It has a diameter of 3.63 megametres.

Neptune's satellite, Triton, has a diameter that is thought to be equal to 3.8 megametres.

STEP 15

10 000 000 metres (10^7 m)
10 megametres (10^1 Mm)

The distance along a meridian of longitude from the North Pole to the Equator is almost exactly 10 megametres.

Actually, this is no coincidence at all. When the metric system was first established, the idea was to pin the basic unit of length to some physical measurement that had no cultural bias in it. It was supposed to be equally meaningful to all nations and cultures.

One thing all humanity has in common is the Earth itself, and the size of the Earth was known. What was wanted, then, was a basic unit of length that was of convenient size, and that was a fixed submultiple of the Earth's circumference. Taking a ten-millionth of the distance from the Pole to the Equator (and therefore a forty-millionth of the circumference of the Earth) would do the trick. The

*A great circle is any circle drawn about the Earth that divides the planet's surface into two equal halves. There are an infinite number of such great circles, but the most familiar are the equator and the meridians of longitude.

metre was established as that distance, and considerable time and effort was spent surveying France and Spain in order to deduce the *exact* distance, so that the metre would be made an *exact* fraction.

Despite all the good intentions, it was a bad idea. Measuring the exact distance along the curvature of the Earth, considering the obstacles in the way, both physical, such as mountains, and human, such as war, was an extremely difficult job. After the figure had been obtained and the metre established, it was inevitable that later measurements would show the earlier ones to have been slightly in error. The officially established metre turned out, after the metric system had been accepted and fixed in detail, to be slightly less than a ten-millionth of the true distance from the pole to the equator.

If the measurement had been exact, the distance from the pole to the equator would have been exactly 10 megametres or 10 000 000 metres. In actual fact, the distance is more like 10.0185 megametres or 10 018 500 metres. That's less than 1/5 of 1 percent off and isn't too bad, but it's not *exact.*

For that reason, the attempt to define the metre in terms of the Earth's size had to be abandoned. Instead, a platinum-iridium rod was kept in an air-conditioned vault in a suburb of Paris, and the distance between two fine scratches on its surface was defined as the metre. This was the distance to which all other "secondary standards" had to conform.

Nowadays, scientists have found a still better way of defining the metre, making use of something physical after all. We'll get to that later on.

At the Step-15 level, we have left the satellites behind us. The largest of the satellites is Ganymede, which circles Jupiter, and it has a diameter of only 5.276 megametres. As a matter of fact, even the smaller planets are behind us now. Mercury has a diameter of 4.878 megametres. Even Mars has a diameter of only 6.796 megametres.

Nevertheless, although in this section of the book we are dealing primarily with straight-line distances, we do include, as "straight line," curved lines that follow a great circle on a spherical surface —otherwise we couldn't talk about airline distances between cities on Earth's surface. Consequently, it is fair to include circumferences of worlds, as well as diameters.

The circumference of Mercury, for instance, is equal to 15.32 megametres, which is comfortably past the Step-15 level. The circumference of Ganymede is 16.58 megametres, and that of Mars is 21.35 megametres.

If we stick with diameters, the smallest planet to exceed the Step-15 level is Venus, with a diameter of 12.1 megametres. The Earth's diameter is 12.756 megametres.

If we follow the curve of the Earth's surface along great circles, here are some distances in the Step-15 range:

New York to Teheran	9.88 megametres
Rio de Janeiro to Berlin	9.99 megametres
Moscow to Capetown	10.10 megametres
Chicago to Tokyo	10.16 megametres
London to Lima	10.16 megametres
Melbourne to New Delhi	10.18 megametres
Caracas to Cairo	10.20 megametres
Mexico City to Rome	10.26 megametres
Montreal to Peking	10.49 megametres
Hong Kong to Madrid	10.55 megametres
Honolulu to Bangkok	10.63 megametres
Paris to Singapore	10.74 megametres

In each of these cases, the distance is equal to that from the pole to the equator and is therefore a fourth of the way around the world.

At the Step-15 range, we can consider distances between planets and their satellites.

Phobos, the inner satellite of Mars, is 9.38 megametres from its planet. This is not much as planet/satellite separations go (indeed, no other satellite we know of is so near the planet it circles). It is not quite the distance from New York to Teheran along a great circle route.

What's more, this separation is measured from the center of Mars to the center of Phobos. From the center of Mars to the surface of Mars is 3.398 megametres, so that, allowing for Phobos's trifling size, too, the distance between Mars and Phobos, *surface to surface,* is just under 6.00 megametres, well below the Step-15 level of length, and only about the airline distance from Montreal to Berlin.

STEP 16

31 600 000 metres ($10^{7.5}$ m)
31.6 megametres ($10^{1.5}$ Mm)

At the Step-16 level we have now reached, we find ourselves beyond distances from city to city on Earth's surface. The distance from England to New Zealand is roughly 20 megametres, and that is halfway around the world on a great circle course. If one tries to go farther than that, the curve of the Earth begins to bring one back to one's starting place. (The distance from the center of Mars to Deimos, the farther of its two satellites, is 20.1 megametres, or about half the distance around the Earth.)

About all that remains to note concerning Earth is its circumference. This is the length about any great circle on the surface of a sphere. If Earth were a perfect sphere, all its infinite number of great circles would be of equal length. Since, thanks to the Earth's revolution upon its axis it is an oblate spheroid, the equator is slightly the longest of the great circles. Its length is 40.074 megametres, and with that we must leave Earth.

There are, of course, four planets that are larger than Earth: Jupiter, Saturn, Uranus, and Neptune. All four present visible surfaces that are not the kind of solid surfaces we are used to in connection with smaller planets, satellites, and asteroids.

What we see, in the case of the four large planets, are the tops of cloud layers. What lies underneath is a subject for speculation, and although we can make some pretty reasonable guesses, we have no directly observed information as yet. Therefore, let us pretend that the visible surface of the cloud layers of these "gas giants" is a surface equivalent to that of Earth, at least for the purposes of computing distances.

The diameter of Neptune, the smallest of the gas giants, is 49.5 megametres, and that of Uranus is 52.3 megametres. Each of these figures is rather larger than that of the circumference of the Earth.

STEP 17

100 000 000 metres (10^8 m)
100 megametres (10^2 Mm)

The Step-17 level of length brings us definitely into the realm of the gas giants. The diameter of Saturn is 120 megametres and that of Jupiter, the largest of the planets, is 143.8 megametres (11 1/4 times the diameter of the Earth). As for the two smaller gas giants, the circumference of Neptune is 155.5 megametres, and that of Uranus is 164.3 megametres.

Uranus has a set of rings circling it that is so dim and sparse as not to be visible on ordinary observation by telescope. The rings were discovered, by accident, in 1977, because they hid, or "occulted," a star in front of which Uranus was passing at the time. The diameter of this ring system from one side of Uranus to the other is about 102.4 megametres, or just about the Step-17 distance.

Of Uranus's satellites, the nearest is Miranda. The distance from the center of Uranus to Miranda is about 130 megametres.

STEP 18

316 000 000 metres ($10^{8.5}$ m)
316 megametres ($10^{2.5}$ Mm)

We are now stretching the giant planets to the uttermost, for the equatorial circumference of Saturn is 377 megametres and that of Jupiter is 451.8 megametres. We must now transfer our attention from the planets themselves to the planetary surroundings.

Jupiter, like Uranus, has sparse rings, not visible by ordinary telescopic examination from Earth. However, Jupiter-probes have detected them. If Jupiter's rings are viewed as continuous objects, the diameter of the outermost edge of the rings is about 260 megametres, a little over half the immense circumference of the planet.

Saturn, though smaller than Jupiter, has the most magnificent set of rings in the Solar system; the only ones that are visible (indeed, spectacularly so) from Earth. The extreme diameter of the Saturnian ring system, as visible from Earth, is 274 megametres or three quarters the circumference of Saturn.

At the Step-18 level of length, we are in the typical range of planet/satellite distances. For instance, the distance from Neptune to its large satellite Triton, surface to surface, is about 330 megametres, which is quite close to the Step-18 level.

That same distance includes three of Uranus's five satellites and nearly reaches the fourth, Titania, which, surface-to-surface, is 412 megametres from Uranus. Dione, Saturn's fourth major satellite, is 317 megametres away from Saturn, surface to surface, and Io, the innermost of Jupiter's giant satellites, is 340.7 megametres from Jupiter, surface to surface.

What's more, the Moon is 378 megametres away from the Earth, surface to surface.

Nor is the Moon the only astronomical object that can approach the Earth at a Step-18 distance. In 1937, an asteroid, perhaps only half a kilometre in diameter, was observed to rush past the Earth at a distance of 780 megametres, twice the distance of the Moon. This Earth-grazer, which approached closer than any other, was named Hermes.

From Hermes' position and motion, its orbit was calculated, and from that orbit it appeared that if Earth and Hermes were both at the points of closest approach of those orbits, Hermes would be about

315 megametres away, only about five sixths the distance of the Moon.

It wouldn't take much of an adjustment of the orbit by the gravitational pulls of Earth and of other planets to place Hermes on a collision course, and even a mere half-kilometre projectile moving at asteroidal velocities could do fearful damage. However, the orbital calculation, based on one flyby, might be inaccurate. Nor have we had the opportunity to correct it, for Hermes has never been seen again.

To be sure, innumerable objects strike Earth every year—indeed, every second. The vast majority are dust-sized particles, and even the spectacular "shooting stars" are caused by bits of matter no larger than pinheads. Only a handful of collisions are by objects large enough to survive passage through the atmosphere and to land as "meteorites." The last potentially damaging strike was in 1908. It took place in central Siberia and, as it happened, killed no one.

Nothing as large as Hermes has struck the Earth through the entire history of civilization.

STEP 19

1 000 000 000 metres (10^9 m)
1 gigametre (10^0 Gm)

At the Step-19 level, we encounter a length of 1 000 000 000 (one billion, by American usage) metres, or 1 000 megametres. We need a new prefix at this stage, and the one adopted in the SI version is "giga-." It is pronounced "JIG-uh" and is symbolized "G." It is from the Latin word "gigas" meaning "giant."

Jupiter's largest satellite, Ganymede, is 1.07 gigametres from the planetary center. (At this distance, it no longer makes much difference whether we calculate from the center of a planet or from its surface.) Beyond Ganymede is Callisto, the outermost of the Galilean satellites; it is 1.88 gigametres from Jupiter.

Saturn's largest satellite, Titan, is 1.22 gigametres from its planet. The satellite beyond it, Hyperion, is 1.48 gigametres from Saturn.

Neptune has a small satellite with an orbit lying beyond that of the large Triton. Whereas most satellites have orbits that are very nearly circular, Neptune's outer satellite, Nereid, has an orbit that is quite elliptical. This means that Nereid comes comparatively close to Neptune at one end of its orbit and then swoops out to a considerably greater distance at the other end. Nereid's closest approach to Neptune is 1.39 gigametres.

Suppose, though, that we consider not just the distance of a satellite from a planet but the length of its orbital path about that planet. For instance, Amalthea, the satellite closest to Jupiter of all those visible from Earth, travels 1.14 gigametres in moving around Jupiter once; that is its orbital length. The orbital length of Mimas, the innermost Saturnian satellite visible from Earth, is 1.16 gigametres.

At this point, it is appropriate to introduce the Sun, which is far larger than any of its planets. Its diameter is 1.39 gigametres, 9 2/3 times that of Jupiter. (If Earth were imagined to be at the center of the Sun, the orbit of the Moon would be little more than halfway to the Solar surface. If Jupiter were at the center, the orbit of Europa would be just below the Solar surface.)

STEP 20

3 160 000 000 metres ($10^{9.5}$ m)
3.16 gigametres ($10^{0.5}$ Gm)

Saturn has a satellite named Iapetus that is next to the outermost. It is at a distance of 3.56 gigametres from Saturn.

As far as the orbital lengths of satellites is concerned, our Moon falls well short of the Step-20 level. The orbital length of the Moon is only 2.41 gigametres.

In comparison, the orbital length of Rhea, the fifth satellite of Saturn (as visible from Earth), is 3.31 gigametres, while that of Oberon, Uranus's outermost satellite, is 3.68 gigametres. The orbital length of Europa, the smallest of Jupiter's giant satellites, is 4.21 gigametres.

The circumference of the Sun is larger still, for it is 4.37 gigametres in length.

STEP 21

10 000 000 000 metres (10^{10} m)
10 gigametres (10^{1} Gm)

Neptune's outer satellite, Nereid, thanks to its highly elliptical orbit, swings out to a distance of 9.73 gigametres at its farthest recession from Neptune.

Saturn's outermost satellite, Phoebe, has a somewhat elliptical orbit. At its closest approach to Saturn, it is 10.83 gigametres from the planet; at its farthest recession, 15.06 gigametres.

Jupiter has four small satellites beyond Callisto, all of which are at a distance of between 11.11 and 11.74 gigametres from Jupiter.

The orbital length of its outermost giant satellite, Callisto, is 11.8 gigametres.

STEP 22

31 600 000 000 metres ($10^{10.5}$ m)
31.6 gigametres ($10^{1.5}$ Gm)

The four outermost satellites of Jupiter are at average distances of from 20.7 to 23.7 gigametres from their planet. Of these, the one with the most eccentric orbit is Pasiphae. At the far end of its orbit, it recedes to a distance of 33.2 gigametres from Jupiter. Of all the known satellites in the Solar system, Pasiphae recedes the farthest from the planet it circles.

The orbital length of Nereid is about 34 gigametres.

At the Step-22 level, we pass beyond the Earth-grazers. The largest and longest known of the Earth-grazers is Eros, and its closest approach to Earth is 22.5 gigametres, while the much smaller Albert, which is located at the extreme range where an object would qualify as an Earth-grazer, does not approach closer than 32.2 gigametres.

We can, in fact, talk about "Sun-grazers." There are occasional comets, known by this name, that approach the Sun at one end of their extraordinarily elongated orbits to distances of ten megametres or less. Individual comets of this type, however, do not appear in our part of the Solar system oftener than once in perhaps many millions of years, and it is therefore hard to talk about them with any familiarity. We can, for that reason, ignore them.

Of the objects that circle the Sun with orbits that are defined throughout, and that make close approaches to the Sun at reasonably short and predictable intervals, that which makes the closest approach is the small asteroid Icarus. At each turn in its orbit (only slightly longer than Earth's), it approaches to within 28.5 gigametres of the Sun.

Mercury, the major planet closest to the Sun, is, at its closest approach, 45.9 gigametres away from it.

And, speaking of the major planets, two of them are occasionally at distances from Earth that are not far above the Step-22 level. While Earth is the third planet from the Sun, Venus is the second, and Mars the fourth. When Venus and Earth are on the same side of the Sun, the closest possible approach is 39.3 gigametres. In the case of Mars the closest approach is 48.7 gigametres.

STEP 23

100 000 000 000 metres (10^{11} m)
100 gigametres (10^2 Gm)

This step brings us to the last gasp of the satellite system. Jupiter's four outermost satellites sweep about the planet in orbits that are 130 to 150 gigametres in length—and there is nothing more we can do with satellites.

We can, however, continue to deal with distances from the Sun. The distance of Venus from the Sun is 108.2 gigametres, while for Earth it is 149.6 gigametres.

Astronomers commonly use the distance from the Earth to the Sun as a convenient measuring unit for planetary distances in general. This distance they call the "astronomical unit," which is symbolized AU. Thus 1 AU = 149.6 Gm.

In the SI version, the astronomical unit is not used, but it is hard to ignore it altogether, since it is so useful in comparisons. Thus, the mean (or average) distance of Mercury from the Sun is 0.387 AU, while that of Venus is 0.723 AU. You can see how much more clearly that relates their distances to that of Earth, than giving the three distances in gigametres would.

STEP 24

316 000 000 000 metres ($10^{11.5}$ m)
316 gigametres ($10^{2.5}$ Gm)

We move beyond Earth now. Mars, the fourth planet, is 227.9 gigametres from the Sun, while Ceres, the largest of the asteroids, is 414.4 gigametres from the Sun.

Thus, The step-24 level of length carries us from the Sun to the asteroid belt and includes what is called the "inner Solar system."

We have entered the region of planetary orbital lengths, too. The orbit of Mercury, the innermost planet, has a length of about 360 gigametres.

STEP 25

1 000 000 000 000 metres (10^{12} m)
1 terametre (10^0 Tm)

At the Step-25 level of length, we have reached the figure of a trillion (American style) metres, and it is time for a new prefix. In this case it is "tera-," which is pronounced "TER-uh" and is symbolized "T." It is from the Greek word "teras" meaning "monster."

The terametre range carries us into the "outer Solar system" as far as planetary distances are concerned. The distance of Jupiter from the Sun falls short, for that planet is only 0.778 terametres from the Sun, but Saturn is 1.427 terametres away.

The length of Earth's orbit about the Sun is 0.94 terametres. This is our final good-bye to the Earth. Nothing that is directly about our planet gives us an example of a greater distance than the length of its orbit about the Sun.

STEP 26

3 160 000 000 000 metres ($10^{12.5}$ m)
3.16 terametres ($10^{0.5}$ Tm)

We are getting toward the outer limits of the planetary system. Uranus, the seventh planet, is 2.870 terametres from the Sun; Neptune, the eighth planet, is 4.497 terametres from the Sun.

Jupiter's orbital length is 4.89 terametres.

STEP 27

10 000 000 000 000 metres (10^{13} m)
10 terametres (10^{1} Tm)

At the Step-27 level of length, we are beyond the most distant known planet, Pluto. Pluto has a mean distance from the Sun of only 5.9 terametres. Of course, its orbit is quite elliptical (more so than that of any other planet). At one end of its orbit, it approaches the Sun just a trifle more closely than Neptune does—and, at the present time and for the next twenty years, it is in that portion of its orbit. At the opposite end of its orbit, however, it is 7.4 terametres from the Sun, still well short of the Step-27 length.

To be sure, the planets are not all there is to the Solar system. There are objects that recede beyond Pluto. This is true of some comets which have long cigar-shaped orbits. Thus, Comet Rigollet is thought to recede to a distance of 8.15 terametres from the Sun, well beyond the farthest point in Pluto's orbit.

That leaves us planetary orbital lengths. Thus, Saturn, in making its mighty sweep about the Sun, travels a length of 8.97 terametres in one revolution.

STEP 28

31 600 000 000 000 metres ($10^{13.5}$ m)
31.6 terametres ($10^{1.5}$ Tm)

The length of Pluto's orbit about the Sun is about 37 terametres. With that, we leave behind simple lengths in the planetary system.

STEP 29

100 000 000 000 000 metres (10^{14} m)
100 terametres (10^{2} Tm)

If we imagine a straight-line distance from the Sun to a point that is 100 terametres away, we will find ourselves far beyond Pluto even at its farthest from the Sun. There is nothing of significance at that distance except for comets, and we have very little in the way of specific statistics for comets that recede from the Sun to such distances. We could use multiple revolutions and say that Pluto in 3 turns about the Sun (or Earth in 115 turns about the Sun) covers a distance of about 100 terametres. However, we can quickly lose ourselves in numbers of revolutions, and we end with something not easily visualized. Let us therefore turn elsewhere.

Light travels through vacuum at a speed of 299 792 500 metres per second. (Reluctantly, I am bringing in a measure of time, which I will be discussing in detail later in the book. Since we all have a pretty good idea what a second is, however, not much is lost, and I'll save the formal explanation for that fundamental unit of time for the proper place.)

By the system we are using to measure distance, we can say that light travels at a rate of just about 300 megametres per second. In one second, then, a traveling beam of light, moving through unobstructed vacuum, will measure off close to a Step-18 distance. In one second, in other words, light will travel from Saturn nearly to its satellite Dione; or from Neptune nearly to its satellite Triton; or, for that matter, from the Earth nearly to the Moon.

A distance of 300 megametres, the distance that light will travel in one second, can be called a "light-second." This is not a unit of distance that is very often used, and it is not permitted in the SI version of the metric system, but it is certainly a dramatic unit.

This way of reckoning distance can be extended. Thus, a "light-minute," the distance light will travel in one minute, would be sixty times as long as a light-second. A light-minute would therefore be 18 gigametres long and would be a little beyond a Step-21 distance. In one minute, light will travel from Jupiter to nearly its outermost known satellite.

It we pass on to a "light-hour," we have a distance which involves another multiplication by sixty. In an hour, a beam of light will travel 1.08 terametres and that is an almost exactly Step-25 distance. In one hour of unobstructed motion through vacuum, light that originates from the Sun will have traveled three fourths of the way to Saturn.

A "light-day" is equal to 24 light-hours, or 25.9 terametres. This is close to a Step-28 distance. If light could be made to follow a gently curved path, it would, in one day, travel nearly completely around Neptune's orbit.

As you see, then, in order to get a notion of a Step-29 distance, where we have nothing conveniently physical to point to, we can use this system and say that it equals 4 light-days. Light, starting at the Sun at a particular moment, will have covered a distance of 100 terametres after an uninterrupted journey of 4 days.

STEP 30

316 000 000 000 000 metres ($10^{14.5}$ m)
316 terametres ($10^{2.5}$ Tm)

In 1973, a comet was detected by the Czech astronomer Lubos Kohoutek as it was approaching the Sun and while it was still as far distant as Jupiter's orbit. For a comet to be visible at such a distance, it seemed fair to suppose that it was a large one and that it would make a spectacular display when it reached the neighborhood of the Earth and the Sun. Unfortunately, it turned out to be a rather rocky comet, one that gave off a far smaller cloud of dust and gas than might have been expected, so that it remained very faint, to the disappointment of all.

Just the same, the long observation of the comet made it possible to work out its orbit with unusual precision for one that was so elongated. The comet's farthest point of recession from the Sun is over sixty times that of Comet Rigollet.

Comet Kohoutek (comets are traditionally named after their discoverers) recedes to a distance of 538.2 terametres. This is substantially greater than a Step-30 distance, and is the greatest securely calculated orbit that involves any specific object of the Solar system.

STEP 31

1 000 000 000 000 000 metres (10^{15} m)
1 petametre (10^{0} Pm)

In the SI system, 1 000 terametres is set equal to 1 petametre. This is pronounced "PEH-tuh-mee-ter" and is symbolized "Pm." A petametre is a quadrillion metres in the American system of naming numbers.

Here it would be fair to talk about a "light-month." The average month is 30.437 times as long as a day, so a light-month would be that many times a light-day. A light-month would therefore be about 0.79 petametres long: not quite a Step-31 distance.

STEP 32

3 160 000 000 000 000 metres ($10^{15.5}$ m)
3.16 petametres ($10^{0.5}$ Pm)

The orbit of Comet Kohoutek is about 2.38 petametres long, or just about 3 light-months long. A beam of light racing along Comet Kohoutek's orbit would take 3 months to complete its journey. This is not quite a Step-32 distance, however, for that is 4 light-months long.

STEP 33

10 000 000 000 000 000 metres (10^{16} m)
10 petametres (10^{1} Pm)

If we continue to work our way up the scale of light-units, we come to the "light-year," the distance that light will travel through unobstructed vacuum in one year of 365.2422 days. The distance it will cover will be 9.46 petametres, or very nearly a Step-33 distance.

In 1950, the Dutch astronomer, Jan H. Oort, suggested there was a shell of comets, a shell containing perhaps as many as 100 000 000 000 of them, slowly turning about the Sun at a distance of 1 to 2 light-years. Occasionally, one of them would be slowed in its passage by collisions or by the gravitational pull of the nearer stars and would drop down toward the inner Solar system. There, in the unaccustomed warmth of the Sun, its icy structure (frozen and undisturbed through all the previous history of the Solar system) would vaporize and form a large cloud of dust and gas, which would become visible to the wondering eyes of human observers.

Through all the long history of the Solar system, only a small fraction of all those comets would have dropped into the inner Solar system. There, eventually, through numerous returns to the vicinity of the Sun, the comet would eventually be disintegrated.

At a Step-33 distance from the Sun, then, assuming Oort's suggestion to be correct, we would be within the shell of comets, a shell that is two thousand times as far from the Sun as Pluto is and eighty thousand times as far from the Sun as Earth is.

STEP 34

31 600 000 000 000 000 metres ($10^{16.5}$ m)
31.6 petametres ($10^{1.5}$ Pm)

If we imagine ourselves moving away from the Sun for a Step-34 distance, 31.6 petametres, we will be beyond every object in the Solar system, even beyond the shell of comets. We will be completely outside the Solar system, and yet we will not have encountered any known object that is not subject to the Sun in the sense of maintaining an orbit about it.

Beyond the Step-34 distance, however, are the stars. It was not till 1838 that a star's distance was worked out for the first time. This was done by the German astronomer, Friedrich W. Bessel, in 1838. The star whose distance was determined by Bessel was one known to astronomers as 61 Cygni, and it is not the closest. The distance of the closest star was published the very next year by the Scottish astronomer, Thomas Henderson.

That nearest star is "Alpha Centauri." It is the brightest star in the constellation "Centaurus," which is so deep in the southern sky that the star is never visible in any but the southernmost portions of the United States.

Alpha Centauri is 41.6 petametres (4.4 light-years) from the Sun, so that it is only a relatively short way beyond the Step-34 distance. It may well be that Alpha Centauri, like our Sun, is surrounded by a shell of comets. Its shell and ours might, in their outermost regions, be within a light-year of each other. In that case, some of the frozen comets circling Alpha Centauri might lie within a Step-34 distance of our Sun.

Alpha Centauri differs from our Sun in one important way. Whereas the Sun is a single star, with no other self-luminous body circling it, Alpha Centauri is a double star. What looks like a single star to the unaided eye is actually two stars separated by a mean distance of about 3.5 gigametres, something more than the distance from the Sun to the planet Uranus. The brighter of the two stars,

"Alpha Centauri A," is almost exactly the twin of our Sun. The dimmer of the two, "Alpha Centauri B," is only a quarter as luminous as either Alpha Centauri A or our Sun.

In 1913, a British astronomer, Robert Innes, discovered a very faint star, not very far away from Alpha Centauri in the sky, and about the same distance from us as Alpha Centauri is. The faint star is a little closer to us, in fact, and is only 40.4 petametres (4.27 light-years) away. It is therefore called "Proxima Centauri," from a Latin word meaning "nearest." It is a small star that circles the far distant Alpha Centauri double star, which is perhaps 1 terametre away from Proxima. Alpha Centauri is therefore a triple-star system, and Proxima Centauri is "Alpha Centauri C."

To summarize: Not far beyond the Step-34 distance, there are three stars that represent our nearest neighbors in space outside our Solar system.

As it happens, astronomers first determined the distances of the nearest stars by measuring their "parallax" (their apparent shift in position, against the background of the farther stars, as the Earth moves from one side of the Sun to the other in the course of a year). This parallax is larger the nearer a star is, but even in the case of Proxima Centauri, where the parallax would be largest, it amounts to only about 0.75 seconds of arc. (A second of arc is so small that it would take 1 296 000 seconds to circle the sky.)

The distance at which a star would have to be located to have a parallax of fully one second of arc is naturally less than the distance of Proxima Centauri. A star would have to be at a distance of 30.84 petametres (3.26 light-years) to have a parallax of one second of arc. Such a distance is called a "parsec" (where "par" stands for "parallax" and "sec" for "second"), and it is almost exactly the Step-34 distance.

Parsec is coming into increasing use by astronomers as an even more convenient measure of distance than light-year, but it is not included among the units permitted in the SI version.

As it happens, there is not a single star within 1 parsec of the Sun.

STEP 35

100 000 000 000 000 000 metres (10^{17} m)
100 petametres (10^2 Pm)

Within 100 petametres of the Sun (3.2 parsecs), there are a dozen stars. These include the three stars of the Alpha Centauri system, of course.

Also included is Sirius, the star that shines most brightly in Earth's sky. It is 81.7 petametres (2.65 parsecs) from us, and is actually a two-star system. The brighter, Sirius A, is larger and more luminous than the Sun. The dimmer, Sirius B, is a tiny star of a type called "white dwarf," a type of star I will have occasion to discuss later in the book.

Just beyond the Step-35 distance, at a distance of 102 petametres (3.3 parsecs), is Epsilon Eridani. This is the nearest star which is Sun-like, in the sense that it is a single star and fairly close to the Sun in size and luminosity.

STEP 36

316 000 000 000 000 000 metres ($10^{17.5}$ m)
316 petametres ($10^{2.5}$ Pm)

With each additional step now, the number of stars included within that distance increases by a factor of about 30. Within 316 petametres (10.25 parsecs) of the Sun, there are perhaps 360 stars.

Included is a double-star system, 61 Cygni, which was the first to have its distance determined. It is 104 petametres (3.37 parsecs) from the Sun. The bright star, Procyon, is 108 petametres (3.50 parsecs) from the Sun.

At a distance of 112 petametres (3.62 parsecs) is Tau Ceti, which is a single star that is even closer in size and luminosity to the Sun than Epsilon Eridani is. When astronomers made their first attempts to investigate the sky for possible radio waves that might indicate the existence of extraterrestrial civilizations, Tau Ceti and Epsilon Eridani were the first stars to which the radio telescopes pointed. (Nothing has been discovered so far.)

Three other bright stars are within Step-36 distances. These are Altair at 157 petametres (5.09 parsecs), Fomalhaut at 216 petametres (6.99 parsecs), and Vega at 247 petametres (8.01 parsecs).

It might be supposed that the reason these stars appear to be so bright is that they are so near to us—and this is certainly an important factor in their brightness. Any star, however luminous, would fade out to invisibility if it were far enough away from us.

Nevertheless, even the closest star would appear really bright only if it were at least as luminous as the Sun is. The farther off it was, the more luminous it would have to be to shine brightly in our skies. The really luminous stars, the ones that are as luminous as the

Sun, or more so, make up only about 1/8 of the total. Seven out of eight stars are dimmer than the Sun, and most of these are considerably dimmer.

These dim stars, small and barely red-hot, are called "red dwarfs" and are not visible to the unaided eye even when they are close to us (as stellar distances go). "Barnard's Star" is a red dwarf and, although it is at a distance of only 55.8 petametres (1.8 parsecs), so that only the Alpha Centauri system is closer, it is so dim that it can only be seen with a telescope.

In fact, Proxima Centauri, the nearest of all the stars, is a feeble red dwarf that is not visible to us without a telescope.

STEP 37

1 000 000 000 000 000 000 metres (10^{18} m)
1 exametre (10^{0} Em)

At the Step-37 distance, we reach 1 000 petametres, which is equal to 1 exametre, or 1 quintillion metres (by American usage). Exametre is pronounced "EK-suh-mee-ter," and it is symbolized as "Em."

Within an exametre of the Sun there are at least 10 000 stars, and some bright ones that we haven't mentioned before are included. These are Arcturus at 0.34 exametres (11 parsecs); Pollux at 0.37 exametres (12 parsecs); Capella at 0.44 exametres (14 parsecs); Aldebaran at 0.49 exametres (16 parsecs); Achernar at 0.61 exametres (20 parsecs), and Canopus at 0.93 exametres (30 parsecs). These are all among the twenty brightest stars in the sky, a group that make up the "first-magnitude stars."

As we move away from the Sun, those stars in the sky that are first-magnitude, despite their distance, must be far more luminous than the Sun. If our Sun were viewed from the distance of Achernar, it would be just barely visible to the unaided eye.

STEP 38

3 160 000 000 000 000 000 metres ($10^{18.5}$ m)
3.16 exametres ($10^{0.5}$ Em)

At Step-38 distances, we include two first-magnitude stars that we had not reached earlier. These are Spica, at a distance of 2.5 exametres (80 parsecs), and Beta Centauri, at a distance of 2.75 exametres (90 parsecs).

STEP 39

10 000 000 000 000 000 000 metres (10^{19} m)
10 exametres (10^{1} Em)

There are no fewer than five bright stars that we have not reached until this step. Alpha Crucis and Antares are each at a distance of 3.7 exametres (120 parsecs), but in different directions, of course; Beta Crucis and Betelgeuse at 4.6 exametres (150 parsecs); and Rigel at 7.7 exametres (250 parsecs).

STEP 40

31 600 000 000 000 000 000 metres ($10^{19.5}$ m)
31.6 exametres ($10^{1.5}$ Em)

Only one first-magnitude star lies beyond the Step-39 distance, and that not by much. It can be included now. It is Deneb, the brightest star in the constellation of Cygnus, the Swan. It is at a distance of 13.25 exametres (430 parsecs).

We have now outstripped the distance of all the bright stars in the sky, and must deal with dim stars only. With this step, in fact, we are leaving what we might consider "the Solar neighborhood."

A distance of 30.84 exametres, which is very nearly the Step-40 distance, is equal to 1 000 parsecs or 1 "kiloparsec." Although this is not a unit permitted in the SI version, it is frequently used by astronomers.

STEP 41

100 000 000 000 000 000 000 metres (10^{20} m)
100 exametres (10^{2} Em)

Stars group themselves in all sorts of ways. There are not only double stars and two or three pairs of double stars in association. There are also large clusters of stars, held in one another's gravitational influence. Among the smaller of these are "open clusters," so-called because they are sufficiently loosely distributed to be open, that is, to have all their stars visible as individuals. Such open clusters may contain anywhere from a score of stars to a thousand of them. They are naturally more easily seen and recognized at a distance as a group, than they would be if their individual stars were randomly distributed over the sky.

There are an open cluster in Sagittarius and three in Auriga that are each at a distance of 37 exametres (1.25 kiloparsecs), and there are other open clusters at twice the distance—76 exametres (2.5 kiloparsecs).

STEP 42

316 000 000 000 000 000 000 metres ($10^{20.5}$ m)
316 exametres ($10^{2.5}$ Em)

There are clusters of stars that are larger and more impressive than the open clusters. These larger clusters are huge spherical arrangements of densely packed stars. There may be hundreds of thousands or even millions of stars in a single "globular cluster" of this sort.

The "Great Globular Cluster" in the constellation Hercules is the most spectacular of these, and it is about 200 exametres (6.8 kiloparsecs) from us.

There are about 100 globular clusters visible in the sky, and it is estimated that there may be an additional hundred that are not visible because they are obscured by dust clouds. The clusters are mostly contained in the constellation of Sagittarius and neighboring regions, and seem to be arranged in a shell about some far distant point.

In fact, when the stars out to the Step-42 distance are studied, it is quite apparent that they are not distributed evenly over space. Instead, they seem to be part of a vast lens-shaped conglomeration of stars, with its center somewhere in the constellation Sagittarius. It is about this center that the globular clusters are arranged, and this center is about 285 exametres (9.2 kiloparsecs) from us.

We cannot see this center in ordinary telescopes because it is obscured by dust clouds, but we know it is there just the same. The shell of globular clusters is one piece of evidence, and, since the advent of radio telescopes, we can detect radio waves (which can penetrate dust clouds) with great ease and precision. The radio waves from the center make the existence and location of that center a virtual certainty.

Because the conglomeration of stars, of which the Sun is part, is lens-shaped, the view of the sky differs according to the direction in which we look. Along the long axis of the lens, we see such a host of very distant and very faint stars that they all melt into a faintly luminous fog, and this is called "the Milky Way." In other directions, we see individual stars without the background of luminous fog.

The lens-shaped conglomeration is therefore called "the Galaxy" from the phrase the Greeks used for the Milky Way.

Now that we have reached the Step-42 distance, then, it becomes useless to consider individual stars. We can only be interested in the Galaxy as a whole.

Suppose, from this point onward, we measure distances from the center of the Galaxy instead of from the Sun. If we considered all the Galaxy that was within 316 exametres of the center, our Sun would be included near the limit of that distance.

There would be stars located at greater distances from the center than our Sun is, but those stars would be relatively few, and sparsely distributed.

STEP 43

1 000 000 000 000 000 000 000 metres (10^{21} m)
1 000 exametres (10^{3} Em)

The SI version of the metric system does not provide for a prefix that will allow us to represent 1 000 exametres (1 sextillion metres) as a unit, and we must therefore remain with exametres to the end.

A Step-43 distance from the center of the Galaxy would encompass the whole of it. The long axis of the Galaxy is 950 exametres (30 kiloparsecs) long, though some recent figures would seem to indicate that it might be only three fourths this long.

Within 1 000 exametres of the Galactic center, there would be up to 300 000 000 000 stars, the large majority being considerably smaller and less luminous than our Sun. What's more, there would be nothing within that distance that was not clearly to be included as part of the Galaxy.

Indeed, up to the 1920s, astronomers suspected that the Galaxy might represent the entire Universe—but let us go on.

STEP 44

3 160 000 000 000 000 000 000 metres ($10^{21.5}$ m)
3 160 exametres ($10^{3.5}$ Em)

If we expand our view to a Step-44 distance from the Galactic center, we find two large star conglomerations, within that distance, that seem to be outside the Galaxy. They are not as large as the Galaxy, but the smaller contains about 50 000 000 stars and the larger about 300 000 000. They might be viewed as "dwarf galaxies."

The nearer of these two is at a distance of 1 500 exametres (50 kiloparsecs), and the farther is at a distance of 1 700 exametres (55 kiloparsecs).

These conglomerations cannot be seen from the North Temperate

Zone, but are visible if one travels southward. The first Europeans to see them, members of Ferdinand Magellan's crew, did so as his ships made their way along the southern coastline of South America in the course of their first-ever circumnavigation of the globe in 1519–1522. They are called the "Magellanic Clouds" for that reason, and they look like detached portions of the Milky Way.

Two other dwarf galaxies exist within the Step-44 distance. There is one in the constellation of Ursa Minor and another in Sculptor. The former is 2 000 exametres (67 kiloparsecs) away, and the latter 2 500 exametres (83 kiloparsecs) away. The Sculptor galaxy has perhaps 5 000 000 stars, and the Ursa Minor galaxy even fewer. Since they are smaller and farther away than the Magellanic Clouds, they cannot be seen in the sky except with a telescope.

A third galaxy is to be found at a distance of 3 100 exametres (100 kiloparsecs), which is very nearly at the Step-44 distance limit.

These five dwarf galaxies might be viewed as satellites of the giant Milky Way Galaxy.

STEP 45

10 000 000 000 000 000 000 000 metres (10^{22} m)
10 000 exametres (10^4 Em)

At this step three more dwarf galaxies are encountered, at distances of 7 100 exametres (230 kiloparsecs), 7 500 exametres (245 kiloparsecs), and 8 500 exametres (275 kiloparsecs).

STEP 46

31 600 000 000 000 000 000 000 metres ($10^{22.5}$ m)
31 600 exametres ($10^{4.5}$ Em)

A distance of 31 600 exametres is equal to 1 025 kiloparsecs or just over 1 megaparsec (a million parsecs). We are thus in the megaparsec range, now.

If we reach out to this distance from the center of the Galaxy, we include, in addition to our Milky Way Galaxy, about twenty dwarf galaxies.

At a distance of 20 000 exametres (0.67 megaparsecs), there is the first giant galaxy we have encountered other than our own. This giant is in the constellation of Andromeda, and it is therefore the "Andromeda galaxy." It is, in fact, about 50 percent larger than our own Galaxy and may have, altogether, some 450 000 000 000 stars.

The Andromeda galaxy is so large and luminous that, despite its vast distance, it is visible to the unaided eye as a small, foggy patch. It, and the much closer Magellanic Clouds, are the only objects outside our own Galaxy that are visible to the unaided eye. The Andromeda galaxy is the farthest object, the very farthest, that the human eye can see without mechanical aid.

At a distance of 30 000 exametres (1 megaparsec), just about the Step-46 limit, there is another large galaxy, not quite as large as the Andromeda galaxy, which is called "Maffei 1," in honor of its discoverer, the Italian astronomer Paolo Maffei. The discovery was not made until the late 1960s because Maffei 1 is seen so near the Milky Way that it is almost entirely obscured by the interstellar dust clouds of the region.

The Milky Way Galaxy, the Andromeda galaxy, and Maffei 1, together with all the dwarf galaxies that cluster between and around them, contain altogether perhaps 1 000 000 000 000 stars, and if that were all there were to the Universe, it would certainly be huge—but that isn't all.

Such is the true size of the Universe that everything within the Step-46 limit (that is, within the megaparsec range) is known simply as "the Local Group." All the galaxies within the Local Group are held together by mutual gravitation and move about like members of a swarm. (At the moment, the Andromeda galaxy is drifting slowly toward us.)

STEP 47

100 000 000 000 000 000 000 000 metres (10^{23} m)
100 000 exametres (10^{5} Em)

As we move outside the Step-46 limit, we leave the Local Group behind us, and from this point on we encounter more and more galaxies, some of them notable for one reason or another.

The galaxy "M81" is so-called because it is the eighty-first item on a list of a hundred fuzzy objects listed by the French astronomer Charles Messier. He was a comet hunter who wanted to make sure that other comet hunters would not mistake these apparently unimportant objects for the fuzziness of a comet. He had no idea that the various objects he listed were enormously more important than comets were.

Galaxy M81 is about 95 000 exametres (3 megaparsecs) away. Like our own Galaxy and the Andromeda galaxy, it exhibits a spiral structure. These three are examples of "spiral galaxies."

Also about 95,000 exametres away is M82, which is an "irregular galaxy," having no clearly defined geometrical structure. (The Magellanic Clouds also are examples of irregular galaxies.) In the case of M82, the irregularity seems to have been imposed on it by a tremendous explosion in its central regions, where the stars cluster most thickly. The explosion has been going on for millions of years, and we are not certain of its cause or nature as yet.

M82 is an example of an "active galaxy," in the sense that it gives off enormous energies as a result of whatever processes are taking place in the center. It is the closest active galaxy to ourselves.

At a distance of 110 000 exametres (3.6 megaparsecs), just beyond the Step-47 distance, is M51, a beautiful spiral galaxy seen at right angles to its galactic plane. The spiral structure is plainly evident —to the extent, in fact, that it has received the name of "the Whirlpool galaxy." It is the first galaxy whose spiral structure was noted, back in 1845, by the Irish astronomer Lord Rosse.

STEP 48

316 000 000 000 000 000 000 000 metres ($10^{23.5}$ *m*)
316 000 exametres ($10^{5.5}$ *Em*)

At a distance of 250 000 exametres (8 megaparsecs) is an irregular galaxy, NGC 4449. It is called that because it is the 4 449th item listed in the "New General Catalog" of galaxies (a listing begun by William Herschel and revised and extended at intervals since then).

The galaxy M104, at a distance of 380 000 exametres (12 megaparsecs), just beyond the Step-48 distance, is a spiral galaxy that is seen edge on. The core bulges upward and downward, while the edge of the spiral cuts across the center. The edge is filled with dark clouds so that it makes a dark stripe across the luminous background. It is called the "Sombrero galaxy" for that reason.

STEP 49

1 000 000 000 000 000 000 000 000 metres (10^{24} *m*)
1 000 000 exametres (10^{6} *Em*)

Now that we have reached Step 49, which is in the septillion-metre range, it becomes difficult to deal with individual galaxies.

Almost all galaxies form part of one cluster or another. Our Local Group is an example of a "galactic cluster," and rather a small one, with no more than two dozen members. There are clusters that are far larger.

For instance, the "Virgo cluster" is a group of at least a thousand galaxies, all gravitationally connected. It is about 500 000 exametres (16 megaparsecs) away, and is the closest giant cluster to ourselves. It has been suggested that it, along with a number of other clusters between itself and the Local Group, all form part of a "local supercluster."

STEP 50

3 160 000 000 000 000 000 000 000 metres ($10^{24.5}$ m)
3 160 000 exametres ($10^{6.5}$ Em)

Another large collection of galaxies is the "Perseus cluster," which is at a distance of about 3 000 000 exametres (100 megaparsecs), or very nearly the Step-50 distance. The largest galaxy in this cluster is NGC 1275. This is a "Seyfert galaxy," so-called in honor of the American astronomer Carl Seyfert, who first noted such objects in 1943.

Seyfert galaxies are a relatively uncommon type of galaxy, marked by a very bright and active central region, one that is much brighter relative to the rest of the galaxy than is the case in ordinary galaxies such as our own. The cause of this activity is not certain, but it seems very likely that violent explosions are going on.

About 3 500 000 exametres (115 megaparsecs) away is another rich cluster, the "Coma cluster," located in the constellation Coma Berenices. This cluster may contain as many as 10 000 galaxies.

STEP 51

10 000 000 000 000 000 000 000 000 metres (10^{25} m)
10 000 000 exametres (10^{7} Em)

As we approach the Step-51 distance, it becomes very difficult to see individual galaxies. A galactic cluster in the constellation Leo is placed at a distance of about 10 000 000 exametres (300 megaparsecs).

In 1963, the Dutch-American astronomer, Maarten Schmidt, showed that certain objects that had been thought of as dim and unremarkable stars of our own Galaxy were actually exceedingly luminous objects at enormous distances. What clearly made them more than ordinary dim stars was the fact that they were the sources of strong radio wave radiation, and that they seemed to be receding from us at huge velocities. Such recession velocities are associated with great distances.

These objects were called "quasi-stellar radio sources," the word "quasi-stellar" meaning "starlike." The phrase was commonly shortened to "quasars." Quasars may be very bright Seyfert galaxies, so distant that only the exceedingly bright pointlike centers of the galaxies remain visible.

From the speed of recession, it is generally assumed that all the quasars are exceedingly distant (although there are astronomers who do not accept this and who try to find ways of showing that the quasars are relatively close to us). If the common view is correct, however, then even the nearest quasar, 3C273 (the 273rd listing in the 3rd Cambridge catalogue of radio sources), is about 10 000 000 exametres away.

STEP 52

31 600 000 000 000 000 000 000 000 metres ($10^{25.5}$ *m)*
31 600 000 exametres ($10^{7.5}$ *Em)*

The Step-52 distance is very close to 1 000 megaparsecs. In other words, we have now reached a distance of 1 gigaparsec (a billion parsecs).

As we approach this distance, there is virtually nothing left we can see but quasars and a few galactic clusters. There is a cluster in the constellation Hydra that may be a little over 28 000 000 exametres (0.92 gigaparsecs) away.

STEP 53

100 000 000 000 000 000 000 000 000 metres (10^{26} *m)*
100 000 000 exametres (10^{8} *Em)*

In 1973, a quasar, OH471, and in 1981, another quasar, 3C427, were found to have speeds of recession so great as to indicate a distance of about 110 000 000 exametres (3.7 gigaparsecs), which is just beyond the Step-53 distance.

These may not only be the farthest objects we have seen so far, but they may be the farthest objects it is possible to see. Beyond them is only an impenetrable haze which may mark the boundaries of the visible Universe (for reasons I will explain later in the book). This boundary may be about 120 000 000 exametres (3.8 gigaparsecs) away.

STEP 54

316 000 000 000 000 000 000 000 000 metres ($10^{26.5}$ *m)*
316 000 000 exametres ($10^{8.5}$ *Em)*

If the distance to the observable edge of the Universe is 120 000 000 exametres, then we may picture the Universe as a globe with a diameter of about 240 000 000 exametres (7.7 gigaparsecs).

STEP 55

1 000 000 000 000 000 000 000 000 000 metres (10^{27} *m)*
1 000 000 000 exametres (10^{9} *Em)*

From what I have said in the previous two steps, the observable Universe can be considered as a globe with a circumference of 750 000 000 exametres, which is only three quarters of the Step-55 distance of 1 octillion metres.

And thus, we run out of Universe. In 54 steps of a half-order of magnitude apiece, we have progressed from the perfectly ordinary distance of 1 metre, which is half the height of a basketball player, to the circumference of the entire Universe.

The Ladder of Length

DOWNWARD

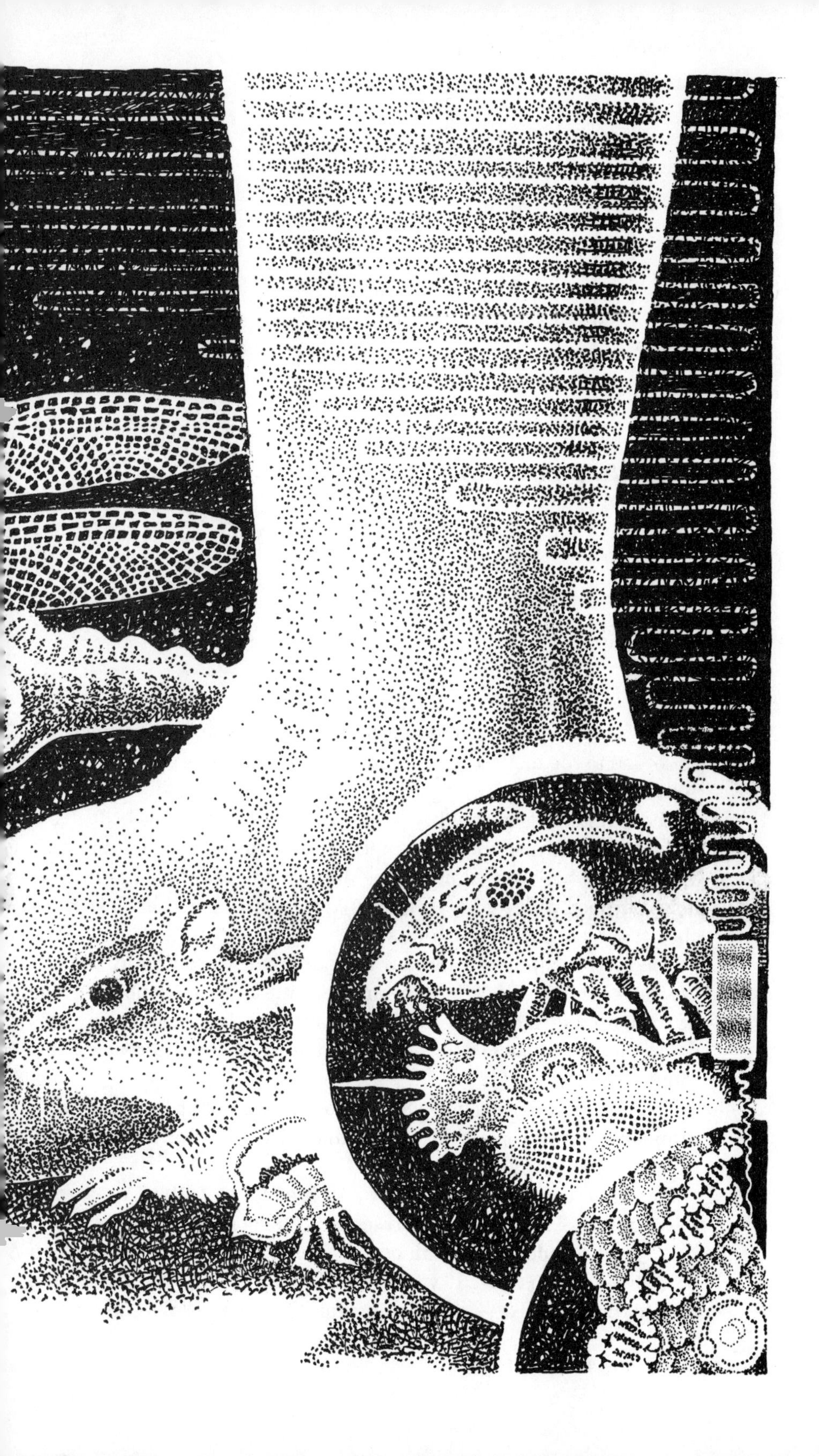

DOWNWARD

STEP 1

1 metre (10^0 m)

Having gone from the metre to the largest reasonable distance, let us return to the metre and remind ourselves that it is a measure that is particularly human.

The human leg is roughly a metre long. The human body from pelvis to head is also roughly a metre long. You can hold a metrestick comfortably between your hands. Many familiar animals are in the neighborhood of a metre in length. So are many artifacts, such as chairs, tables, bookcases, and so on.

The metre is indeed a human measure, but we can't stay with it. We have multiplied it before, and we must divide it now. As we headed up the ladder of length toward ever-increasing distances before, so we must now head down it toward ever-decreasing distances.

The question arises, however, as to how we ought to represent a number less than 1 (either an ordinary fraction or a decimal fraction) in exponential figures.

For instance, 1 metre divided by 10 is equal to 1/10 of a metre or to 0.1 metre. How is that represented in exponential figures?

We have decided that 1 can be represented as 10^0, and that 10 can be represented as 10^1. If we say "1 divided by 10," we are saying "10^0 divided by 10^1," and asking for the answer. By the rule of subtracting exponents in division, which I explained near the beginning of the book, we have decided that "10^0 divided by 10^1" is 10^{0-1} or 10^{-}. We conclude that 1/10 (which we might also write as $1/10^1$) is equal to 10^{-1}.

We can easily show in the same manner that $1/100 = 1/10^2 = 10^{-2}$; that $1/1\,000 = 1/10^3 = 10^{-3}$, and so on. In general, $1/10^n = 10^{-n}$. (We can read such numbers as "ten to the minus first power," "ten to the minus second power," and so on; or, more briefly, as "ten to the minus one," "ten to the minus two," and so on.)

In such negative exponentials, the numeral in the exponent represents the number of zeroes in the denominator of the number when it is expressed as an ordinary fraction. Thus, since the fraction 1/10 000 000 has seven zeroes in the denominator, we know that it is 10^{-7}. And if we want to express 10^{-12} as an ordinary fraction, we count zeroes and write 1/1 000 000 000 000.

Since 1/10 is the same as 0.1, we know that $0.1 = 10^{-1}$ also. Similarly, 0.01 is 10^{-2}; 0.001 is 10^{-3}, and so on. As you see, the numerical portion of the exponent is equal to the number of zeroes in the decimal fraction, provided you remember to place, and count, one zero to the left of the decimal point. Thus 10^{-28} may be written as 0.000 000 000 000 000 000 000 000 000 1. (It makes much more sense to use the exponential number, as I think you will agree.)

In taking steps downward, we will do as we did in the steps upward. Instead of multiplying by 3.16 each time, however, we will divide by 3.16. If we divide 1 metre by 3.16, we will get 0.316 455 6 . . . and we will just call that 0.316. Then if we divide 0.316 by 3.16, we will get 0.1, and so on.

In this way, we will go down the ladder in steps of half an order of magnitude, from 10^0 to $10^{-0.5}$ to 10^{-1}, and so on.

STEP 2

0.316 metres ($10^{-0.5}$ m)
3.16 decimetres ($10^{0.5}$ dm)

The prefix "deci-" (DESS-ih) was adopted by the original devisers of the metric system to represent a tenth of a basic unit, so that a "decimetre" (DESS-ih-mee-ter) is a tenth of a metre.

The prefix is from the Latin word for "tenth," and it is unfortunate that it is so similar to "deka-," which is used for a measure that is ten times the basic unit, so that a dekametre is equal to 100 decimetres. (The confusion is rendered even worse, when dekametre is spelled—as it sometimes is—decametre.) Since "deci-" is symbolized as "d," and "decimetre," therefore, as "dm," therefore, "deka-" must be symbolized as "da" and "dekametre" as "dam."

Fortunately, "deci-" is not often used these days, and "deka-" is even more rarely used, or the similarity might be unbearable.

Since a metre is equal to 39.37 inches in common American measure, 0.316 metres is equal to 12.44 inches, or just a trifle over a "foot."

The foot does not belie its name; it originated as the length of the human foot. Legend persists in supposing that it was originally the length of some distinguished foot such as that of Charlemagne. That might be. We would expect the length of the foot to be greater for a tall man than a short man, and my foot (I am of average height) is 11 1/4 inches long, whereas Charlemagne, who was a six-footer, may well have had one 12 inches long.

Common small animals are in this range in height or length: cats, dogs, badgers, rabbits, chickens, ducks, and so on.

There are animals in this range which belong to groups we usually think of as larger. For instance, we think of deer as moderately large animals. The largest deer is the Alaskan moose, which stands 2 metres tall at the shoulder. The smallest deer, however, is the mouse deer or "Lesser Malayan chevrotain," which may be not more than 2 decimetres (8 inches) high at the shoulder. It is the smallest ruminant (cud-chewing animal) in existence.

Even more extreme is the case of the dinosaurs, usually thought of as mountainous animals, and indeed, as a group, they were the largest land animals that ever lived. Nevertheless, the smallest known dinosaur, *Compsognathus,* was about the size of a chicken, and not more than 6 decimetres long.

There are also objects in this range that belong to classes we usually think of as smaller. For instance, we think of eggs as comparatively small objects, certainly considerably shorter than a foot. Nevertheless, on the island of Madagascar, there once lived a bird called the Aepyornis, the heaviest bird that ever lived. It may not have become extinct till 1660, and may have been the source of the legendary "roc" or "rukh," the impossibly large bird featured in the tales of Sinbad the Sailor in *The Arabian Nights.* The egg laid by the Aepyornis is the largest egg ever laid by any creature, and it may have approached 3 decimetres in length.

Then again, we think of insects as small animals indeed, and far below the Step-2 dimension. Yet there are exceptions. There are species of stick insects that are unusually long, the longest having been measured at 3.3 decimetres (13 inches). Other insects are large by virtue of their wings—as, for instance, butterflies. There is a giant butterfly in the Solomon Islands with a wingspan of just over 3 decimetres. For that matter, there were prehistoric dragonflies—extinct now for 200 000 000 years at least—that were twice as large and had a wingspread of up to 7 decimetres (27 1/2 inches). This means that the largest insects possessed a longer measurement than the smallest dinosaur, an unusual overlapping indeed.

A sound wave which is 3.16 decimetres in length will, in pitch, be the "do" above "high do," a fairly shrill sound.

An electromagnetic wave which is 3.16 decimetres in length is in the shortest section of the radio wave region. For this reason, it is said to be in the "microwave" region. "Micro-" is from the Greek "mikros" meaning "small" and is usually used by scientists for things that are *very* small. Thus, "microorganisms" are living things too small to be seen, and a "microscope" is a device for making such things visible. Microwaves, however, are not *very* short. Indeed, there are electromagnetic waves far, far shorter than microwaves, and the only significance of the prefix in this case is to indicate that the waves are the smallest of the radio wave region. "Microwave" is therefore rather a misnomer.

Microwaves became particularly important during World War II, when they came to be used in radar.

The atmosphere is opaque to radio waves generally, because there are regions rich in electrically charged particles in the upper atmosphere (those regions are referred to as the "ionosphere") which reflect radio waves. This means that radio waves originating on Earth cannot pass through the atmosphere into outer space, and that radio waves originating in outer space cannot pass through the atmosphere to Earth's surface.

Microwaves, however, *can* get through the ionosphere, and from the surface of the Earth, we can detect these very short radio waves when emitted by objects in space. The study of these waves is now called "radio astronomy" and is carried on by means of "radio telescopes."

STEP 3

0.1 metres (10^{-1}m)
1 decimetre (10^{0} dm)

A decimetre is about equal to 4 inches in common American measure. The width of the average male hand (including the thumb) is about 1 decimetre, and "hand" is an old unit of measure equal to 4 inches in length. It is still used today as a measure of the height of horses at the shoulder. They are said to be so many hands high.

There are animals at the Step-3 range of length. A chipmunk is about 1.5 decimetres long, excluding its tail. The smallest species of mouse can be up to 1.35 decimetres long, *including* its tail.

There are also some beetles that may not be as long as some stick insects but that are far bulkier, and that extend into this range. The Goliath beetle can be almost 1.5 decimetres long, and can thus vie in size with a tailless chipmunk.

A sound wave that is 1 decimetre long is very nearly the equivalent in pitch to the highest note on the piano keyboard.

An electromagnetic wave 1 decimetre long is still in the microwave region. An electromagnetic wave 2.1 decimetres in length is given off by cold hydrogen undergoing a certain atomic change.

Since hydrogen is the most common substance in the Universe, and since cold hydrogen makes up, in particular, most of the thinly spread-out matter in interstellar space, it is reasoned that *any* intelligent species, anywhere in the Universe, would want to study that particular portion of the microwave spectrum. With everyone studying it, it would seem that if any intelligent species wanted to send out a signal, they would send it out at or near that wavelength, knowing that other intelligent species would be equipped to receive it. For that reason, astronomers have inspected various parts of the sky with a particular eye out for radiation of this type. They have not found anything of apparent intelligent origin yet.

STEP 4

0.031 6 metres ($10^{-1.5}$ m)
3.16 centimetres ($10^{0.5}$ cm)

The prefix "centi-" (SEN-tih), symbolized as "c," represents a hundredth of a basic unit, from the Latin "centum" meaning "hundred." A "centimetre," therefore, is a hundredth of a metre. The prefix is not commonly used, except in "centimetre," and its use is falling off even there.

Here we leave behind the smallest adult warm-blooded creatures. The smallest mammal is the pygmy shrew, which is between 4 and 5 centimetres long, excluding the tail. The smallest bird is the bee hummingbird, which is about the same length as the pygmy shrew.

We can't expect warm-blooded animals to be any smaller than this, for the body temperature is maintained by the quantity of living tissue in the body, while heat is lost through the surface. The smaller the creature, the greater the amount of surface compared to the quantity of living tissue. It becomes harder and harder, therefore, as size decreases, to maintain heat production at a rate equal to heat loss, so that body temperature remains normal. As it is, shrews and hummingbirds must eat continually and are constantly on the brink of starvation.

Of course, newborn animals are much smaller than adults. For instance, the ordinary chicken egg we have for breakfast is about 5 centimetres long, and that is therefore the size of a newborn chick. Bird eggs, however, cannot possibly maintain their heat of themselves, but must be kept warm from the body heat of an adult bird, or from some other outside source.

A sound wave that is 3.16 centimetres long is approaching the limit of shrillness that the human ear can hear, and an electromagnetic wave that is 3.16 centimetres long is approaching the lower wavelength limit of the region useful in radar.

STEP 5

0.01 metres (10^{-2} m)
1 centimetre (10^{0} cm)

A centimetre is equal to just about 2/5 inch. The smallest bird egg known is that of the bee hummingbird and is not more than about 1.15 centimetres long. It is the smallest bit of warm-blooded life that exists.

Cold-blooded animals can be smaller than any mammal or bird, since there isn't the constraint of having to have sufficient tissue to produce the heat required to maintain a temperature higher than that of the environment.

The tiniest reptiles are species of geckos with lengths (excluding the tail) of only 1.8 centimetres. There are species of frogs in Cuba that, even in adulthood, attain a length of 1.25 centimetres at most, and are about the size of bee hummingbird eggs. The smallest fish, a pygmy goby found in lakes and rivers in the Philippines, is a bit under a centimetre in length and is thought to be the smallest vertebrate creature in existence.

All these are, of coure, multicellular organisms, each one built up of, and containing, numerous cells, some highly specialized.

To be sure, an egg, however large—even an Aepyornis egg—is, strictly speaking, a single cell, but there are adult organisms that are also single cells and these are "unicellular organisms." These were first discovered in 1676 by the Dutch microscopist, Anton van Leeuwenhoek.

Unicellular organisms are generally considerably smaller than multicellular animals. There is, however, as in almost all classifications on the basis of size, an overlap. The largest living unicellular organism is a protozoon (a free-living animal cell) which can have a length of up to 1.5 centimetres, thus matching, in this respect, the smallest vertebrates, even though the latter are multicellular. (The Nummulites, now extinct, were protozoan cells with a length of up to 2.4 centimetres.)

A sound wave that is 1 centimetre long is too shrill for any human ear to hear. It is in the "ultrasonic region," as is any sound wave that is shorter still, so we now leave the sound waves behind as offering nothing further of interest.

An electromagnetic wave that is 1 centimetre long is beyond the radar region. Waves of about this length are emitted by molecules (groups of atoms) and, when detected on Earth as issuing from interstellar clouds, can be used to identify the nature of molecules in space.

STEP 6

0.003 16 metres ($1^{-2.5}$ m)
3.16 millimetres ($10^{0.5}$ mm)

The prefix "milli" (MIL-ih), from the Latin "mille," meaning "thousand," is symbolized as "m," just as "metre" is. A millimetre is therefore symbolized as "mm." Increasingly "milli-" is replacing "centi-" and "deci-" in use. We are approaching the point where 1 centimetre will routinely be referred to as 10 millimetres, and 1 decimetre as 100 millimetres. This is even more true where these prefixes are used for any basic measure other than "metre."

A millimetre is a thousandth of a metre. Therefore, 3.16 millimetres are equal to about 1/8 inch in common American measure.

At this stage, we are beginning to edge toward the lower limit of even those familiar creatures that are unusually small. The smallest centipede is only 4.8 millimetres long. On the other hand, the largest known insect eggs, those of a species of moth, are about 3.2 millimetres long.

STEP 7

0.001 metres (10^{-3} m)
1 millimetre (10^{0} mm)

One millimetre, which is about 1/25 inch, is getting toward the lower limit of what we can comfortably make out with unaided eyes.

The smallest spider known, one found in Australia, is about 0.8 millimetres long. At least the female is; the males are only 0.6 millimetres long. The smallest ants are of about the same size.

Electromagnetic waves that are 1 millimetre in length stand at the lower limit of the radio wave range. Those waves that are shorter still are called "infrared waves." This is a purely artificial and arbitrary dividing point, representing no fundamental change in properties.

STEP 8

0.000 316 metres ($10^{-3.5}$ m)
316 micrometres ($10^{2.5}$ μm)

The prefix "milli-" was the one used to represent the smallest fraction of a basic unit, as established by the original designers of the metric system. In contemporary times, however, a series of prefixes has been devised for every three orders of magnitude that we move downward on the scale of dimension.

Thus the prefix "micro-" (MIKE-roh), which I mentioned earlier in connection with microwaves, represents a millionth of a basic unit. A "micrometre" (MIKE-roh-mee-ter), therefore, is a millionth of a metre, or a thousandth of a millimetre.

The prefix "micro-" is an unfortunate one in that its initial letter "m" is also that of "milli-." Since "milli-" has preempted "m" as a symbol, and "mega-" has preempted "M," "micro-" is forced to use as its symbol the Greek alphabet equivalent of "m," which is "mu." That letter, in the Greek alphabet, is written μ, so that "micrometre" is symbolized as "μm."

Before the use of "micrometre" was standardized in the SI version of the metric system, the term "micron" had come into common use as a millionth of a metre, and still is met up with on occasion—but it is on no account to be used in the SI version.

The smallest insects we know of fall just below the Step-8 distance and are only about 200 micrometres long. Such tiny insects are no bigger than sand grains of typical size.

Electromagnetic waves that are 316 micrometres long are still in the infrared region.

STEP 9

0.000 1 metres (10^{-4} m)
100 micrometres (10^{2} μm)

The smallest multicellular organisms belong to the class Rotifera. There are 2 000 species of them, and even the largest are not over 500 micrometres in length. The smallest are only about 80 micrometres in length. Naturally, although such organisms consist of more than one cell, they do not consist of many cells—just a few.

The largest cell in the human body is the ovum, or egg cell, produced by the female. It is large because it contains a food supply for the developing embryo in its very earliest stage, before it is affixed to the wall of the uterus and begins obtaining its food by diffusion, across the placental membrane, from its mother's blood stream. The ovum is 140 micrometres across, so that it is larger than a small rotifer.

The smallest plant seeds, those of certain orchids, are perhaps not more than 100 micrometres across—as are the smallest insect eggs. In contrast, the smallest grains of sand are about 60 micrometres across.

STEP 10

0.000 031 6 metres ($10^{-4.5}$ m)
31.6 micrometres ($10^{1.5}$ μm)

Step 10 brings us into the range of ordinary one-celled organisms. These fall into three groups: protozoa, which are related to animal cells; algae, which are related to plant cells; and bacteria, which, strictly speaking, are neither.

Bacteria are, in general, smaller than either protozoa or algae, and are, in fact, the smallest cells there are, but, again, there is an overlap. The largest bacterial cells have lengths of up to 45 micrometres.

STEP 11

0.000 01 metres (10^{-5} m)
10 micrometres (10^{1} μm)

The common paramecium is one of the most familiar of the protozoa, and is often viewed by students who are looking through a microscope for the first time. They are about 10 micrometres long, and can just be seen as moving specks in water under a strong light. Ordinary cells of the human body are also in this range. The human liver cell is about 12 micrometres across.

STEP 12

0.000 003 16 metres ($10^{-5.5}$ *m*)
3.16 micrometres ($10^{0.5}$ μ*m*)

Most of the infrared region of electromagnetic waves cannot pass through the atmosphere. Infrared wavelengths in the Step-12 range and shorter, however, can pass through the atmosphere. Much of the infrared in sunlight is in this range of wavelength, and can reach the Earth's surface.

Many bacterial cells are in the Step-12 size range, with lengths of 3 micrometres or so.

The familiar red blood corpuscles (which are not complete cells because they lack a cell nucleus) are small discs about 7.7 micrometres across the long diameter, and are only about 3.7 micrometres thick.

STEP 13

0.000 001 metres (10^{-6} *m*)
1 micrometre (10^{0} μ*m*)

The smallest cell in the human body is the sperm cell, produced by the adult male. It consists of a head, which, in turn, is made up of little more than a cell nucleus—half a cell nucleus, in fact—plus a long tail, the lashing of which propels it through a watery medium. It contains the hereditary material (chromosomes), which is added to the ovum on fertilization. The ovum contains another half-nucleus, contributed by the mother, with its own equal supply of chromosomes, plus an initial food supply. The head of the sperm cell is about 2 micrometres long.

An *E. coli* cell, the ordinary bacterium of the human large intestine, is also about 2 micrometres long.

The cells do not have homogenized contents, but contain discrete "organelles" which perform certain functions and which are, of course, smaller than the cells themselves. The nucleus, which contains the cell's reproductive mechanism, may be, typically, 2 micrometres across. A mitochondrion, which is the cell's energy-producing organelle, may be 1.5 micrometres across.

STEP 14

0.000 000 316 metres ($10^{-6.5}$ m)
316 nanometres ($10^{2.5}$ nm)

The prefix "nano-" (NAN-uh) is from the Greek word "nanos," meaning "dwarf." It is symbolized as "n" and represents one-billionth (by American usage) of the basic measure, so that the "nanometre" is a billionth of a metre, or a thousandth of a micrometre. Before the establishment of the SI version of the metric system, a nanometre was often referred to as a "millimicron," and this is still encountered at times, but that usage is now forbidden.

The smallest bacterial cells are some that were discovered in sewage in 1936. These, called pleuropneumonia-like organisms (PPLO) have, at their largest, diameters of 300 nanometres.

These PPLO bacteria are the smallest free-living cells there are, that is, the smallest cells that contain all the machinery for life.

There are, however, cells that do not contain all the machinery for life, but must grow and reproduce, parasitically, inside complete cells, making use of some of the host's machinery for the parasite's own purposes.

The largest of these parasitic bodies are called "rickettsia" because they were discovered in 1906 by the American physician, Howard Taylor Ricketts. Rickettsia are generally disease-producing and, in human beings, cause such diseases as typhus and Rocky Mountain spotted fever.

Rickettsia may be incomplete, but they tend to be larger than the free-living PPLO. Some rickettsia are as large as 800 nanometres across, though others are as small as 200 nanometres across.

There are also viruses, first detected by the Dutch botanist Martinus W. Beijerinck in 1898. The viruses are also incomplete and can grow and reproduce only inside complete cells. They cause a great many human diseases: chicken pox, smallpox, mumps, measles, poliomyelitis, influenza, the common cold, and so on.

On the whole, viruses are even smaller than rickettsia, but the largest viruses, those which cause cowpox or tobacco mosaic disease, for instance, are 300 nanometres across. Thus, viruses overlap rickettsia, and both overlap true bacteria.

As far as electromagnetic waves are concerned, the infrared region ends at 760 nanometres. At wavelengths as short as that and shorter, the retina of the eye is affected, and we see the electromagnetic waves as "light."

What's more, we see different colors depending on the wavelengths. The colors merge into each other very gradually as the wavelengths shorten, so that it is meaningless to try to set up a boundary line between them.

However, at 700 nanometres, you have a clear red color; at 610, orange; at 575, yellow; at 525, green; at 470, blue; and at 415, violet.

Red is at the long-wavelength end of the "spectrum" (as the band of visible light was named by Isaac Newton, who, in 1666, was the first to produce and study it). It has the longest waves of any form of visible light and, therefore, the lowest frequency—that is, the fewest number of waves formed in one second. As one moves to waves longer than 760 nanometres, the red color (very deep and dim by then) fades out. That is why the still longer waves, with still lower frequencies, are called "infrared" (below the red). The frequencies are below those associated with red light.

At the other end of the spectrum, there is violet, which fades out as the wavelengths grow smaller and the frequency higher. At 360 nanometres, the retina is no longer affected, and the light vanishes. Beyond that point are "ultraviolet" rays (beyond the violet) with shorter wavelengths and higher frequencies than violet.

This band of visible-light wavelengths can pass through the atmosphere easily, and what's more, it is in just this band that the radiation from the Sun reaches its peak. It was very useful to have life-forms, generally, develop sense organs that respond to this section of the total range of electromagnetic radiation. It not only reaches us, but it is the richest portion of the whole.

The section of ultraviolet radiation with wavelengths from 380 nanometres down to 300 nanometres can get through the atmosphere. From 300 nanometres downward, however, the atmosphere is opaque again.

When light waves were studied by a Swedish physicist, Anders Jonas Ångström, in 1868, he used a unit that was a ten-billionth of a metre (by American usage). In 1905, such a unit was named an "angstrom unit" and symbolized as Å. The wavelengths of visible light were said to stretch from 3 800 Å to 7 600 Å. Such a unit is still often met with.

An angstrom unit is a tenth of a nanometre, and in the SI version of the metric system, angstrom units are out of bounds. Nanometres are used instead, and the wavelengths of visible light stretch from 380 nanometres to 760 nanometres.

There is a rare gas named "krypton," which exists in several varieties, one of which is krypton-86. When krypton-86 is heated by an electric current, it glows with an orange color that is due to its emission of light of a wavelength of about 605.78 nanometres. This is the only wavelength that is emitted, and its length can be measured very accurately. Ever since 1960, that wavelength has been used as the standard for determining the exact length of the metre. The metre is defined as equal to 1 650 763.73 wavelengths of the orange light emitted by krypton-86 under closely specified conditions.

STEP 15

0.000 000 1 metres (10^{-7} m)
100 nanometres (10^{2} nm)

The smallest of the PPLO organisms, when first formed, has a diameter of 100 nanometres, and that is the smallest free-living form of life known to exist.

The influenza virus has a diameter of 115 nanometres, and would seem to be a tiny sphere.

The chromosomes in cell nuclei carry the factors of heredity, and each unit factor is called a "gene." There are some estimates that a typical gene is about 125 nanometres long.

All living things possess giant molecules that have no sign of life about them at all, in or out of cells. The most important of these are proteins, and one of the larger protein molecules is about 160 nanometres across.

STEP 16

0.000 000 031 6 metres ($10^{-7.5}$ m)
31.6 nanometres ($10^{1.5}$ nm)

We now approach the lower limit of life. Cells, bacteria, and rickettsia have been left behind, and all that are left are viruses. A bacteriophage that parasitizes the small bacterial cell of *E. coli* is itself only about 25 nanometres across, and the smallest recognized virus is about 20 nanometres across. That is the smallest bit of life, the smallest scrap of matter that can reproduce itself.

STEP 17

0.000 000 01 metres (10^{-8} m)
10 nanometres (10^{1} nm)

We can still have something to do with life at the Step-17 level, if we take particularly small measurements. The tobacco mosaic virus is rodlike in shape. The length of the rod, as I said before, is about 300 nanometres, but the width is only 15 nanometres.

There are also particularly small organelles within cells. Ribosomes are the organelles within which protein molecules are manufactured. The ribosomes within the *E. coli* bacteria are about 18 nanometres across.

STEP 18

0.000 000 003 16 metres ($10^{-8.5}$ m)
3.16 nanometres $10^{0.5}$ nm)

At this level, we must deal with nonliving molecules: proteins. In mammalian red cells, for instance, there are molecules of a protein called "hemoglobin," which pick up oxygen when the red cells pass through the tiny blood vessels that line the lungs. In doing so, the hemoglobin is converted to oxyhemoglobin. The oxyhemoglobin gives up the oxygen when passing through the tiny blood vessels that line and penetrate all the other tissues of the body. In this way, oxygen is carried from the atmosphere to the cells of the body, and life can continue. The individual molecule of hemoglobin is 6.8 nanometres across.

The hemoglobin molecule is a quadruple molecule, actually. The molecule is made up of four chains of atoms that each include an iron atom. A similar molecule in muscle is made up of a single chain with an iron atom. That is "myoglobin" and its diameter is 3.6 nanometres.

STEP 19

0.000 000 001 metres (10^{-9} m)
1 nanometre (10^{0} nm)

Electromagnetic waves with a length of 1 nanometre represent an arbitrary lower wavelength-limit to the ultraviolet region. Beyond the 1-nanometre limit are the X rays, first discovered by the German physicist Wilhelm Roentgen in 1895.

In the case of molecules, we must go on to molecules smaller than those of ordinary protein molecules. The molecules of sucrose, the ordinary sugar we add to our coffee, is about 1 nanometre across.

STEP 20

0.000 000 000 316 metres ($10^{-9.5}$ m)
316 picometres ($10^{2.5}$ pm)

The prefix "pico-" (PEEK-uh) from a Spanish word signifying "a small amount" represents a trillionth (using the American system) of a basic unit, and is symbolized as "p." A "picometre" (PEEK-uh-mee-ter) is a trillionth of a metre, and is symbolized as "pm." An angstrom unit is equal to 100 picometres.

We have now reached the realm of very small molecules. A molecule of glucose, the simple sugar found in grapes and in blood, is 700 picometres across. A molecule of alanine, one of the simpler amino acids, chains of which make up the protein molecules, is only 500 picometres across. A molecule of carbon dioxide is 300 picometres across.

Molecules are built up of atoms. Glucose consists of twenty-four atoms (six carbon, twelve hydrogen, and six oxygen); alanine consists of thirteen atoms (three carbon, seven hydrogen, two oxygen, and a nitrogen); carbon dioxide consists of only three atoms (one carbon and two oxygen).

Some of the larger atoms are also to be found, individually, in this range. A cesium atom is over 500 picometres across, and a mercury atom is about 300 picometres across.

STEP 21

0.000 000 000 1 metres (10^{-10} *m*)
100 picometres (10^{2} *pm*)

At the Step-21 level we are at the smallest atoms: hydrogen, carbon, nitrogen, oxygen, and so on, all of which are roughly 100 picometres across.

STEP 22

0.000 000 000 031 6 metres ($10^{-10.5}$ *m*)
31.6 picometres ($10^{1.5}$ *pm*)

We have now left the atoms behind. Where electromagnetic waves are concerned, those with a wavelength of 31.6 picometres are still in the X-ray range.

STEP 23

0.000 000 000 01 metres (10^{-11} *m*)
10 picometres (10^{1} *pm*)

Here, more or less arbitrarily once more, we can set up a boundary line between X rays and a group of radiation of still shorter wavelength, the gamma rays. Gamma rays are all the electromagnetic waves with wavelengths of less than 10 picometres.

STEP 24

0.000 000 000 003 16 metres ($10^{-11.5}$ m)
3.16 picometres ($10^{0.5}$ pm)

The atom is nuclear in nature. That is, it possesses an "atomic nucleus" at its center, a nucleus that is small in comparison to the atom as a whole. This nucleus is surrounded by a number of light particles called electrons. The number of electrons varies from 1 to over a hundred depending on the particular atom-variety, or "element," involved. These electrons are arranged (to put it simplistically) in a number of shells, the number varying from one to seven depending on the number of electrons. The greater the number of shells, the closer to the nucleus the innermost shells are.

If we imagine ourselves beginning at the center of the nucleus and reaching out 3.16 picometres in all directions, we reach beyond the nucleus and the inner shells of electrons, but do not reach the outer shells.

STEP 25

0.000 000 000 001 metres (10^{-12} m)
1 picometre (10^{0} pm)

We are still shrinking toward the atomic nucleus. At a distance of 1 picometre from it, only the innermost shells of the larger atoms are liable to be included.

STEP 26

0.000 000 000 000 316 metres ($10^{-12.5}$ m)
316 femtometres ($10^{2.5}$ fm)

The prefix "femto" (FEM-tuh) represents a quadrillionth (by the American system) of a basic measure and is symbolized "f." A "femtometre" (FEM-tuh-mee-ter) is a quadrillionth of a metre and is symbolized "fm." The prefix is from a Danish word "femten" meaning "fifteen" since a femtometre is 10^{-15} metres. Before the prefixes of the SI version of the metric system were established, a femtometre was sometimes referred to as a "fermi," after the Italian physicist, Enrico Fermi, who did important work in nuclear physics. That has been eliminated now, however.

At this step, we are at last in the range of the atomic nucleus. The largest atomic nuclei have diameters of something like 630 femtometres, and are made up of about 250 small "subatomic particles" called "neutrons" and "protons."

STEP 27

0.000 000 000 000 1 metres (10^{-13} *m*)
100 femtometres (10^{2} *fm*)

The individual proton and neutron have diameters of about 100 femtometres, and with that we come to a halt. There are particles that, in some ways, are smaller than the proton and neutron, and there are some distances involved in our understanding of the Universe that are smaller than the diameter of these particles, but it will be more convenient to take up such matters in other categories of ladders of the universe.

As it is, in the downward ladder of distance, we have traveled from an ordinary body-sized distance to the smallest particles that can be considered to have a diameter in the ordinary sense of the word, and we have done it in 26 steps of half an order of magnitude each.

In fact, if we combine the upward ladder of distance with the downward one, we go from the diameter of the proton at the lowest extreme, to the circumference of the Universe at the upper extreme, and we do it in 80 steps, or 40 orders of magnitude.

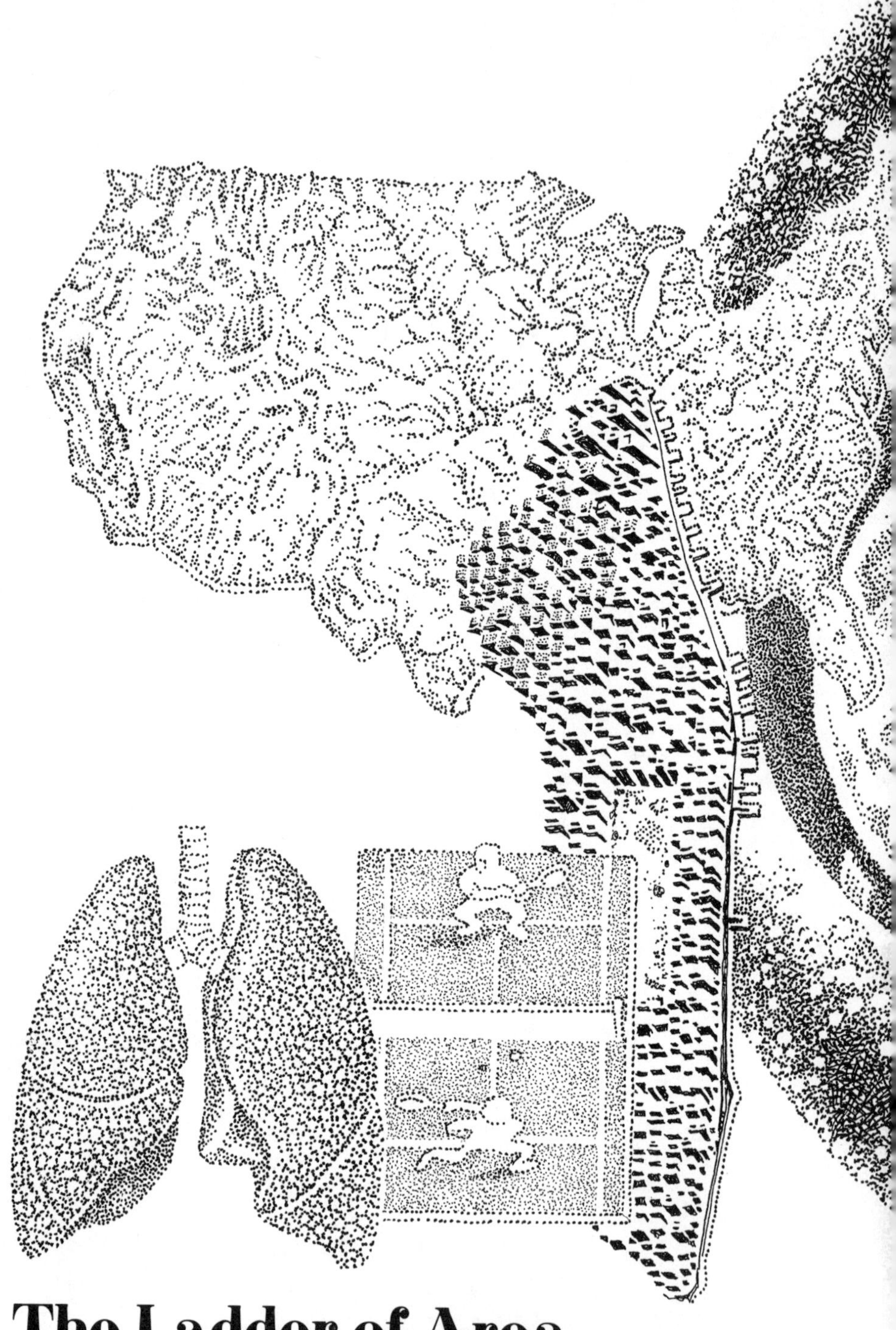

The Ladder of Area

UPWARD

R. Jones

UPWARD

STEP 1

1 square metre (10^0 m^2)

The area of a flat ("two-dimensional") figure of any shape, or of the surface of a solid ("three-dimensional") figure, is calculated by multiplying one length by another.

Suppose you have a square—a flat, four-sided figure—where all four sides meet at right angles and are of equal length. If each side is equal to 2 metres, then the area is 2 metres × 2 metres, or 4 square metres (with "square metre" symbolized as "m^2").

The areas of figures that are not squares can be worked out in other, more complicated ways, but invariably it boils down to the product of a length by another length, and the answer (if the lengths are in metres) is always in square metres.

The simplest situation is, of course, to have a square with each side equal to 1 metre. The area is 1 metre × 1 metre, or 1 square metre. The square metre is the basic measure of area in the SI version of the metric system, and we can start with that.

Any figure can have an area of 1 square metre, of course. An equilateral triangle will have an area of 1 square metre if each of its three equal sides has a length of about 1.52 metres. The surface of a sphere will be 1 square metre if the diameter of the sphere is about 0.564 metres in length.

If we wish to bring the square metre closer to home, then consider that a tall, lean man has a surface area close to 2 square metres. (Naturally, calculating the area of something as irregular and complicated as the surface of a human body is much more difficult than working it out for regular geometric figures, but it can be done.)

In common American measure, the nearest thing to a square metre is a "square yard." A square in which each side is 1 yard long has an area of 1 square yard. Since a yard is equal to 0.9144 metres, a square yard which is 1 yard $\times$ 1 yard is equal to 0.9144 metres $\times$ 0.9144 metres or 0.8361 square metres (or, roughly, 5/6 of a square metre). This is the same as saying that 1 square metre is equal to 1.196 square yards (or, roughly, 1 1/5 square yards).

It may seem annoying to have to remember this conversion, and the best thing is not to remember it at all, but simply to get into the habit of using square metres.

STEP 2

10 square metres (10^1 m^2)

In the ladder of length, we worked with half-orders of magnitude so that while Step 1 was 1 metre, Step 2 was 3.16 metres, and so on. In the ladders of area, we are working with a length multiplied by a length. Although $1 \times 1 = 1$, $3.16 \times 3.16 = 10$. In other words, while the lengths go up by half an order of magnitude, the product of those lengths, the areas, goes up by a full order of magnitude.

Therefore, in the ladder of area, we are going to move up an order of magnitude at a time. Having begun with 1 square metre at Step 1, we will go to 10 square metres at Step 2, then to 100 square metres, and so on. Each step will be ten times the size of the previous step.

A typical middle-class bedroom in a New York apartment (mine, for instance) has a floor area of about 10 square metres.

STEP 3

100 square metres (10^2 m^2)
1 square dekametre (10^0 dam^2)

Imagine a square, each side of which is 1 dekametre long. The area of the square is 1 dekametre $\times$ 1 dekametre = 1 square dekametre (and a square dekametre is symbolized "dam^2"). A dekametre is, however, equal to 10 metres. We can equally well calculate the area of the square, therefore, as follows: 10 metres $\times$ 10 metres = 100 square metres.

It follows that while 1 dekametre is equal to 10 metres, 1 square dekametre is equal to 100 square metres. When a metric prefix increases a smaller measure of length by one order of magnitude it increases a smaller measure of area by two orders of magnitude. If it increases a smaller measure of length by three orders of magnitude it increases a smaller measure of area by six orders of magnitude, and so on.

A square dekametre is often spoken of as an "are" (pronounced just like the English word, and symbolized as "a"). This is frequently used as a unit of area in the metric system and can even be used in the SI version.

In thinking about area, we must remember that objects of the same general size need not always have the same area. An object may, for instance, be smooth and have a relatively small area; or it may be wrinkled and corrugated, and although it is of the same general size, it will then have a much greater surface area.

The human lungs, for instance, are relatively small, air-filled bags. Their inner surfaces are, however, divided into several hundred million tiny little bags called "vacuoles," each with a very small surface area. Add all the areas, however, and the sum is much higher than if the lungs were bags with a single, smooth inner surface.

The total inner surface of the human lungs is estimated to be some 85 square metres. Compare this with a tennis court for singles, which is about 195 square metres in area. We've each got nearly half a tennis court of surface inside our lungs, which accounts for the efficiency with which we absorb oxygen. We could not maintain life if we had lungs with a smooth inner surface.

A similar situation exists for our brain, where it is the cells on the surface that do the thinking. The brain is wrinkled ("convoluted") so that there is more surface to it than if it were smooth. We are more intelligent than other animals not only because our brain is larger, but also because it is more convoluted. (Yet dolphins, which are our size, have brains that are not only larger than ours, but more convoluted as well. It makes one wonder.)

STEP 4

1 000 square metres (10^3 m^2)
10 square dekametres (10^2 dam^2)

The entire playing area of a football field from the 50-yard line to the 27-yard line in either direction is just about 10 square dekametres. This amounts to a little less than a fourth of a football field. Again, 10 square dekametres is about equal in area to four tennis fields of the size used for playing doubles.

An archaic measure of area, still very commonly used in the United States, is the "acre." Originally, it signified the amount of land that could be plowed by a yoke of oxen in one day, and the word comes from an old term for "field." (The Latin word for "field" is "ager," and that gives us "agriculture," for instance.)

An acre is 160 square rods, where the "rod" is a measure of length equal to 5.5 yards. Each square rod is therefore 5.5 yards times 5.5 yards, or 30.25 square yards. An acre is equal, then, to 160 times 30.25, or 4 840 square yards.

This means that an acre is equal to 4 047 square metres, or a little over 40 square dekametres. The Step-4 area of 10 square dekametres is thus equal to about a quarter of an acre.

Needless to say, acres are not part of the SI version of the metric system.

STEP 5

10 000 square metres (10^4 m^2)
1 square hectometre (10^0 hm^2)

Since a hectometre is equal to 10 dekametres, or to 100 metres, a square hectometre is equal to 100 square dekametres, or to 10 000 square metres.

Since a square hectometre is equal to 100 square dekametres—that is, 100 ares—a square hectometre can be referred to as a "hectare" (remember that "hecto-" is the metric prefix signifying a hundred times the indicated measure, and that in this case the "o" of the prefix is omitted in the interest of euphony). "Hectare" is pronounced "HEK-tahr" and is symbolized as "ha." The hectare, in places using the metric system, is even more commonly used as a measure of area than the are is.

A square hectometre, or hectare, is nearly twice the area of a football field and is equal to nearly 2.5 acres.

The block on which my apartment house building stands is a rectangle. It is 213 metres long east and west; 80 metres long north and south. The area of the block is, therefore, 213 metres times 80 metres, or 17 040 square metres. This is 1.7 square hectometres (or nearly seven acres).

STEP 6

100 000 square metres (10^5 m^2)
10 square hectometres (10^1 hm^2)

We have now reached the level of notable islands. For 51 years, from 1892 to 1943, the major immigration station for foreigners entering the United States with intent to live there permanently was Ellis Island. This is in New York Bay, 1.7 kilometres southwest of Manhattan Island. Millions entered, and among them, in February, 1923, was a 3-year-old boy named Isaac Asimov.

The area of Ellis Island is 11 hectares (or 27 acres).

STEP 7

1 000 000 square metres (10^6 m^2)
1 square kilometre (10^0 km^2)

One square kilometre is equal to 100 square hectometres, or 100 hectares (or 247 acres). The square kilometre is the common measure of geographic area in the world. In the United States alone, the common measure of area is the "square mile." Since a kilometre is equal to 0.6214 miles, a square kilometre is equal to 0.6214 miles times 0.6214 miles, or 0.3861 square miles. In other words, a square kilometre is equal to just over 3/8 of a square mile, while a square mile is equal to nearly 2 2/3 square kilometres. Another way of putting it is that 8 square kilometres is just about equal to 3 square miles. The best way of putting it, however, is to learn to deal with square kilometres and forget the square miles.

The smallest internationally recognized independent state in the world is Vatican City, which is headed by the Pope, and which has a population of about 1 000. It is 44 hectares in area, just four times the size of Ellis Island, or 0.44 square kilometres. It is on the western bank of the Tiber River and is located entirely within the city of Rome.

Second in smallness of size is Monaco, on the Mediterranean shore, near the southeastern tip of France. Its population is 25 000, and it has an area of 1.5 square kilometres, which is 3 1/2 times that of Vatican City.

Central Park, in Manhattan, has an area of 3.4 square kilometres. It is therefore 7.7 times the area of Vatican City and 2 1/4 times the area of Monaco. (When the American actress, Grace Kelly, married Prince Rainier III of Monaco back in 1956, such was the overblown romance of the whole in the American newspapers, that the Prince was treated as though he were the ruler of a realm of immense size. When I told my friends that Monaco was less than half the size of Central Park, their expressions were rather stronger than doubtful. I think they suspected my sanity.)

STEP 8

10 000 000 square metres (10^7 m^2)
10 square kilometres (10^1 km^2)

We are still in the realm of small geographic regions.

Gibraltar is not an independent nation; it is a peninsula off the southern shore of Spain, near the Strait of Gibraltar, and is one of the last colonial remnants of the recently vast British Empire. It was taken by the British in 1704 during the War of the Spanish Succession and has been held by them ever since. It has a population of 30 000 and is about 5.8 square kilometres in area, or 1.7 times the area of Central Park.

On the upper side of Step 8 is Macao, a last remnant of the recently large Portuguese Empire. It is located in Canton harbor in China, just across the bay from Hong Kong, and has been Portuguese since 1557. It has a population of about 250 000 and an area of 15 square kilometres, 4 1/2 times that of Central Park.

STEP 9

100 000 000 square metres (10^8 m^2)
100 square kilometres (10^2 km^2)

The island of Manhattan, with a population of 1 500 000, is 57 square kilometres in area, nearly four times the area of Macao and about 16 3/4 times that of Central Park. It is the smallest of the five boroughs that make up New York City and, in area, falls well short of the Step-9 stage.

The next smallest borough, the Bronx, also with a population of 1 500 000, is just at the Step-9 stage, with an area of 109 square kilometres, twice that of Manhattan.

Among independent regions there is San Marino, a small state in central Italy near the Adriatic Sea. It has preserved its independence now for some 15 centuries and may be the oldest republic in the world. It has a population of 20 000 and an area of 61 square kilometres, just a trifle above that of Manhattan.

Liechtenstein is an independent principality between Austria and Switzerland, with a population of 25 000 and an area of 157 square kilometres, three times that of Manhattan.

STEP 10

1 000 000 000 square metres (10^9 m^2)
1 000 square kilometres (10^3 km^2)

All five boroughs of New York City, taken together, have a population of about 7 000 000 and an area of 945 square kilometres, right at the Step-10 range.

New York City is the most populous city in the United States, but it is not the largest in area. Among the other cities a million-plus in population, Los Angeles, with nearly 3 000 000 people, has an area of 1 202 square kilometres; and Houston, with about 1 500 000 people, has an area of 1 401 square kilometres.

Outside the United States, there is Hong Kong, another remnant of the British Empire. Located across Canton harbor from Macao, it has a population of 4.5 million and an area of 1 045 square kilometres. It is only slightly larger than New York City.

At this stage, we must leave cities behind, but we can greet the astronomical bodies. These come in all sizes, and there are vast numbers that are small enough to have areas far below a square metre. Of those bodies that are important enough to have names, however, there is that brick-shaped, Earth-grazing asteroid, Eros, which has a surface area of about 550 square kilometres, half that of Los Angeles. There is also Phobos, the larger of the two satellites of Mars. Phobos has a surface area of about 1 500 square kilometres, or just a bit more than the area of Houston.

STEP 11

10 000 000 000 square metres (10^{10} m^2)
10 000 square kilometres (10^4 km^2)

In the United States, we have entered the realm of the states and territories. We are well past Rhode Island, the smallest state, with a population of 950 000 and an area of 3 140 square kilometres (about twice that of the city of Houston); and Delaware, with a population of 600 000 and an area of 5 325 square kilometres.

We must move on to Connecticut, the third smallest, with a population of 3 100 000 and an area of 12 970 square kilometres, which is on the upper side of the Step-11 stage. Puerto Rico, with a population of 3 300 000 and an area of 8 890 square kilometres, is on the lower side.

Abroad, there is the independent island of Cyprus in the eastern Mediterranean (its northeastern third was seized by Turkey in 1974). It has a population of 635 000 and an area of 9 251 square kilometres, three times that of Rhode Island.

The independent island of Jamaica, in the West Indies, has a population of 2 300 000 and an area of 11 422 square kilometres, almost that of Connecticut. Lebanon, which is on the east-Mediterranean shore of Asia, just north of Israel, has an area of 10 360 square kilometres.

STEP 12

100 000 000 000 square metres (10^{11} m^2)
100 000 square kilometres (10^5 km^2)

We are in the region of middle-sized states now. Ohio, with a population of 10 800 000, has an area of 106 720 square kilometres; while, among independent nations, Iceland, with a population of 235 000, has an area of 102 950 square kilometres. Both are close to the Step-12 level.

In outer space, we are in the range of the surface area of small (but not tiny) satellites. Himalia, the largest of the outer satellites of Jupiter, has a surface area of 90 790 square kilometres, which is just about equal to the area of the European nation of Hungary.

STEP 13

1 000 000 000 000 square metres (10^{12} m^2)
1 square megametre (10^0 Mm^2)

Since a megametre is equal to 1 000 kilometres, a square megametre is equal to 1 000 000 square kilometres; just as 1 square kilometre is equal to 1 000 000 square metres.

At the square-megametre (Step-13) level, the geographic areas have become large. Texas itself falls short. With a population of 14 200 000, it has an area of but 0.692 square megametres. Alaska, the largest American state, has a population of 400 000 and an area of 1.518 square megametres. It is the only state above the square-megametre level.

Abroad, there is Egypt which, with a population of 43 000 000 and an area of 1.000 square megametres, hits the level virtually exactly.

In space, Enceladus, one of the smaller satellites of Saturn, has a surface area of 0.785 square megametres, which is a little larger than the area of Texas. Pallas, the second largest asteroid, has a surface area of 0.985 square megametres, very nearly equal to the area of Egypt.

STEP 14

10 000 000 000 000 square metres (10^{13} m^2)
10 square megametres (10^1 Mm^2)

At this stage, nothing but the largest nations will serve. The United States, with a population of 230 000 000, has an area of 9.36 square megametres, which falls slightly short of the Step-14 level. China, with a population of nearly 1 000 000 000 and an area of 9.6 square megametres, and Canada, with a population of 24 000 000 and an area of 9.97 megametres, come closer. Only the Soviet Union, with a population of 268 000 000 and an area of 22.3 square megametres, lies above (and well above) the Step-14 level.

As for outer space, Ceres, the largest asteroid, falls far short, for its surface area is only about 3.5 square megametres, only three eighths the area of the United States.

Still, we must not underestimate the asteroids.

Astronomers suggest that there may be as many as 100 000 asteroids altogether. Suppose there are and that they are each 4 kilometres in diameter. There would be as much material in those 100 000 asteroids as there would be in a single asteroid with a diameter of 185 kilometres. The single asteroid with a 185-kilometre diameter would have a surface area equal to 107 000 square kilometres. The hundred thousand asteroids, with diameters of 4 kilometres each, would each have a surface area of 50 square kilometres; so that all together they would have a surface area of 5 000 000 square kilometres, or 5 square megametres.

In other words, the same amount of material, divided into comparatively small bits, has far more surface area than it would have had as a single object. It is the same principle of divisions and subdivisions that gives the human lungs such an enormous inner surface. It is for this reason that space enthusiasts sometimes wonder if the asteroids might not someday represent a major living space for human beings.

For a *single* astronomical object with a surface area approaching the Step-14 level, there is Rhea, the second largest of Saturn's satellites. Its surface area is 7.35 square megametres, about equal to the area of Australia.

STEP 15

100 000 000 000 000 square metres (10^{14} m^2)
100 square megametres (10^2 Mm^2)

Even whole continents cannot reach the Step-15 level of area. Asia, the largest continent, has an area of 44 square megametres, which is twice that of the Soviet Union, but is less than half the Step-15 level.

Asia, Africa, and Europe, together, make up the "World Island." They are the largest chunk of land in the world, of which no part is totally separated from any other by a sizable stretch of water.

Asia and Europe are attached all along the line of the Urals, and are considered as separate continents only through the accident that civilization began in the eastern Mediterranean where, owing to a limited horizon, Asia and Europe seemed separated by the Aegean and Black Seas. "Eurasia" is the better geographical term. As for Africa, it is attached to Asia by a landbridge at Suez.

The World Island has an area of just about 85 square megametres. Add to it Australia and all the islands off the shores of the World Island, and you have some 4 100 000 000 people spread over an area of very nearly 100 square megametres. In short, it takes all the land of the Eastern Hemisphere to reach the Step-15 level.

The ocean surface far surpasses the land surface on Earth. The ocean, in fact, makes up 70 percent of the total surface of the Earth. The Atlantic Ocean, for instance, with an area of 82 square megametres, is very nearly equal in area to the World Island—and it is only the second largest ocean.

The Pacific Ocean, which is largest, has a surface area of 181 square megametres, and is considerably larger, in itself, than all the land area on Earth. It is the only single feature of the Earth's surface that betters the Step-15 level of area.

In space, we are beyond the satellites. The total surface area of our Moon is merely 38 megametres, only seven eighths that of Asia. The largest satellite, Ganymede, has a surface area of 87 megametres, which slightly surpasses that of the Atlantic Ocean.

Mars, however, the second smallest planet, has a total surface area of 145 square megametres, about four fifths that of the Pacific Ocean, and well above the Step-15 level.

STEP 16

1 000 000 000 000 000 square metres (10^{15} m^2)
1 000 square megametres (10^3 Mm^2)

At this level, when we have reached an area in the range of a quadrillion square metres, we find we have left our entire planet behind. The Earth has, in total, a surface area of 520 square megametres, or little more than half the Step-16 level.

Venus, a slightly smaller world, has a surface area of 460 square megametres. Add the Moon to these two planets, and the total surface area of these three bodies is 1 018 square megametres, and in this way, the Step-16 level of area is reached.

STEP 17

10 000 000 000 000 000 square metres (10^{16} m^2)
10 000 square megametres (10^4 Mm^2)

The total surface area of all these smaller planets, plus that of all the satellites and asteroids, taken together, may not much surpass 2 000 square megametres, and we must turn to the four planetary giants.

These do not have a surface area in the same way that the smaller bodies of the Solar system have; that is, they do not have a solid surface that we know of, and, if they do, we have no idea of its size. What we see as the surface is actually the top of a cloud layer, and we may, if we wish, consider the surface area of that cloud layer.

The total surface area of Neptune (the smallest of the giant planets) is 7 700 square megametres, and that of Uranus is 8 600 square megametres. With these, the Step-17 level is approached.

STEP 18

100 000 000 000 000 000 square metres (10^{17} m^2)
100 000 square megametres (10^5 Mm^2)

Now no planet is sufficient. Even Jupiter, the largest planet, has a total surface area (along its cloud layer) of 65 000 square megametres, well below the Step-18 level of area.

However, the total surface area of the two largest planets, Jupiter and Saturn, taken together, is 110 000 square megametres. Add to this the surface of all the known remaining nonluminous bodies of the Solar system, and the figure may move up to 130 000 square megametres. With the Step-18 level, then, we run out of the Solar system (excluding the Sun), and if we are to move on to higher levels, we must move beyond the planets, and reach for the stars.

Some stars are considerably smaller than large planets, and such "condensed stars" will be dealt with later in the book. Among the "normal" stars, most are smaller than the Sun. The smallest stars are "red dwarfs," and the star Luyten 726-8 B is about as small as a red dwarf can get. Its diameter may be as little as 180,000 kilometres and its surface area may be about 100 000 square megametres, which is just the Step-18 level of area.

STEP 19

1 000 000 000 000 000 000 square metres (10^{18} m^2)
1 square gigametre (10^0 Gm^2)

A million square megametres is equal to 1 square gigametre, which is equal to a quintillion square metres. Here we pass beyond the red dwarfs. We must move on to stars that are beyond the red dwarfs but are nevertheless still smaller than the Sun.

Alpha Centauri B, the dimmer of the two stars making up Alpha Centauri, has a diameter of about 973 megametres and a surface area of 3 square gigametres, which puts it well above the Step-19 level of area.

STEP 20

10 000 000 000 000 000 000 square metres (10^{19} m^2)
10 square gigametres (10^1 Gm^2)

At the Step-20 level of area, we have passed the Sun, which, with a diameter of 1.4 gigametres, has a surface area of 6 square gigametres (nearly 50 times the total surface area of all the other sizable bodies of the Solar system, taken together).

To reach the Step-20 level, we must move on to those stars that are larger than the Sun. Procyon has a diameter of 1.7 gigametres and its surface area is just about 9 square gigametres.

STEP 21

100 000 000 000 000 000 000 square metres (10^{20} m^2)
100 square gigametres (10^2 Gm^2)

A still larger star, Achernar, has a diameter of 5.5 gigametres and a surface area of 95 square gigametres, nearly 16 times that of our Sun, and very nearly at the Step-21 level of area.

STEP 22

1 000 000 000 000 000 000 000 square metres (10^{21} m^2)
1 000 square gigametres (10^3 Gm^2)

At the Step-22 level, we are speaking of a sextillion square metres and are beginning to approach the upper level of normal stars. The largest of these have diameters of as much as 28 gigametres and surface areas of about 2 500 square gigametres, substantially above the Step-22 level, and over 400 times the surface area of our Sun.

STEP 23

10 000 000 000 000 000 000 000 square metres (10^{22} m^2)
10 000 square gigametres (10^4 Gm^2)

Now we must move on to the abnormally large stars, the "red giants," which, like the condensed stars that are abnormally small, occur late in the evolutionary life cycle of stars. The bright star, Arcturus, one of the smaller red giants, with a diameter of 37 gigametres, has an area of 17 000 square gigametres.

STEP 24

100 000 000 000 000 000 000 000 square metres (10^{23} m^2)
100 000 square gigametres (10^5 Gm^2)

The red giant Beta Pegasi has a diameter of 150 gigametres. If it existed in place of the Sun, its bloated structure would place its surface out beyond the orbit of Mercury. Its surface area would be 280 000 square gigametres, well beyond the Step-24 level of area.

STEP 25

1 000 000 000 000 000 000 000 000 square metres (10^{24} m^2)
1 square terametre (10^0 Tm^2)

A million square gigametres is equal to a square terametre (and to a septillion square metres).

Antares, the bright red star of the constellation Scorpio, is a large red giant with a diameter of 550 gigametres. Were it in place of the Sun, its surface would be located beyond the orbit of Mars. Its surface area is 3.8 square terametres, substantially above the Step-25 level of area.

STEP 26

10 000 000 000 000 000 000 000 000 square metres (10^{25} m^2)
10 square terametres (10^1 Tm^2)

We are approaching the largest of the red giants. Betelgeuse, the bright red star of the constellation Orion, pulsates. Its diameter expands at times till it is 1.1 terametres in width. In the place of the Sun, the surface of Betelgeuse would lie about halfway between Jupiter and Saturn when the star was fully extended. Its surface area would then be 15 square terametres.

STEP 27

100 000 000 000 000 000 000 000 000 square metres (10^{26} m^2)
100 square terametres (10^2 Tm^2)

The largest red giant whose diameter has been estimated is Epsilon Aurigae B. The diameter is given as 2.8 terametres, nearly three times that of Betelgeuse at its largest. In place of the Sun, its surface would lie at the orbit of Uranus, and its surface area of 100 square terametres would be at just the Step-27 level of area.

In 26 steps, then, covering 26 orders of magnitude, we have gone from a square metre, half the surface area of the body of a tall, lean adult male human being, to the immense surface area of the largest red giant.

To go further would require us to add the surface areas of the various stars in a globular cluster, a galaxy, a galactic cluster, and the Universe. A rough estimate might lead us to suppose that the total surface area of all the stars in the Universe might be somewhere about 10^{40} square metres, which would be equal to ten trillion stars the size of Epsilon Aurigae B. It would not be worthwhile, however, to work our way through thirteen more orders of magnitude, since we could do so past this level more usefully when we work up the ladder of volume and of mass later in the book.

The Ladder of Area

DOWNWARD

R. JONES

DOWNWARD

STEP 1

1 square metre (10^0 m^2)

Let us now return to the square metre in preparation for moving down the ladder of area for as many steps as will prove useful, taking one order of magnitude per step, as we did on the way upward.

A square metre is a unit that seems to be snugly within the human range. The desk I work at, for instance, has a rectangular surface which, since it was manufactured in the United States, has measurements that work out evenly in common American units. It is 30 inches by 60 inches. That is 0.762 metres by 1.524 metres, so that it has a surface area of 1.16 square metres.

The table at which my wife and I eat has a circular surface that has a radius of 25 inches (0.635 metres). That means it has a surface area of 1.27 square metres.

The standard mattress on which a single person can sleep comfortably in a middle-class American home is 36 by 75 inches, which represents an area of 1.75 square metres.

An area of between one and two square metres is thus a convenient size on which the basic activities of life can be carried out.

STEP 2

0.1 square metre (10^{-1} m^2)
10 square decimetres (10^1 dm^2)

By now, we are undoubtedly accustomed to the manner in which metric units fit each other in round numbers. We expect to find 1 square metre equal to 100 square decimetres, so that 10 square decimetres is equal to 0.1 square metre.

In common American units, on the other hand, there are 3 feet to a yard and 12 inches to a foot. This means that a square in which each side is 1 yard long has also a side which is 3 feet long. The area of this square is either 1 yard × 1 yard, or 3 feet × 3 feet, so that it is either 1 square yard or 9 square feet.

In other words, if 1 yard is equal to 3 feet, 1 square yard is equal to 3 × 3, or 9 square feet. By similar reasoning, 1 square foot is equal to 12 × 12 or 144 square inches, and 1 square yard is equal to 12 × 3 × 12 × 3, or 1 296 square inches.

Consider! How many Americans can tell you, offhand, that there are 144 square inches to a square foot, or 1 296 square inches to a square yard, or 9 square feet to a square yard. These figures are virtually never remembered, even if they have been learned in school.

When Americans therefore object to the metric system as "strange" and say they prefer the "familiar" measurements they are used to, they are not really aware of the true situation. The so-called familiar measurements are unfamiliar to nearly everyone, and are bound to be so, because they are uneven and irregular. That is one of the chief reasons why the common measurements should be replaced by the metric system.

As it happens, a square yard is equal to 0.8361 square metre, and since a square foot is equal to 1/9 square yard, it is equal to 0.0929 square metres, or to 9.29 square decimetres. In other words, the Step-2 measure of area, working downward, places us right in the square-foot range.

An ordinary sheet of typewriting paper in the United States is 8 1/2 by 11 inches. It has an area of just about 6 square decimetres. The ordinary manila envelope used for paper of this size is 9 by 12 inches, or just about 7 square decimetres in area. The manila envelope of the next larger size, as commonly sold in stationery stores, is 10 by 13 inches, or 8.4 square decimetres in area. I have two checkbooks, the larger of which is 9 3/4 by 14 inches, or 8.8 square decimetres in area. My unabridged dictionary lying open on my desk is 12 by 22 inches, or 17 square decimetres in area.

These familiar objects straddle the Step-2 (downward) measure of area.

STEP 3

0.01 square metres (10^{-2} m^2)
1 square decimetre (10^0 dm^2)

The checks I make out in paying my bills are of two kinds. Those made out on behalf of the corporation of which I am president (Nightfall, Inc.) are rectangles that are 3 by 8 1/4 inches in size and that have an area of just about 1.6 square decimetres. My personal checks, however, are 2 3/4 by 6 inches, and are 1.06 square decimetres in area.

The ordinary envelopes I use in personal mail are 3 5/8 by 6 1/2 inches, or about 1.5 square decimetres in area. The postcards sold by the American post office are 3 1/2 by 5 1/2 inches, or 1.24 square decimetres in area.

The blank cards I used in preparing indexes to my books are 3 by 5 inches, or 0.97 square decimetres in area.

STEP 4

0.001 square metres (10^{-3} m^2)
10 square centimetres (10^1 cm^2)

At this level, we are approaching the square-inch range. An inch is equal to 2.54 centimetres, so a square inch is equal to 2.54×2.54, or 6.45 square centimetres. Therefore, 10 square centimetres is equal to 1.55 square inches.

The 40-cent airmail stamp currently sold by the United States Post Office, the one picturing Philip Mazzei (an Italian-American who served as an American agent in Europe during the Revolutionary war), is a rectangle 1 by 1 3/5 inches in size, with an area of almost exactly 10 square centimetres.

The ordinary 20-cent stamp used for local mail within North America has an area of about 5.5 square centimetres, while the gummed return address labels I affix to my letters have an area of about 6.6 square centimetres (or just a trifle over 1 square inch).

STEP 5

0.000 1 square metres (10^{-4} m^2)
1 square centimetre (10^0 cm^2)

The ordinary paper clips I use are 3 centimetres long and 0.7 centimetres wide. The area marked out by the curving bit of wire is just about 2 square centimetres, therefore.

The ball-point pen I most frequently use is very nearly 1 centimetre wide at its widest. If I were to cut it in two at its widest point, the area of the circular cross section would be just under 0.8 square centimetres.

STEP 6

0.000 01 square metres (10^{-5} m^2)
10 square millimetres (10^1 mm^2)

A three-letter word as typed by my typewriter or as printed out by my word processor—a word such as "can"—would take up an area of about 10 square millimetres.

STEP 7

0.000 001 square metres (10^{-6} m^2)
1 square millimetre (10^0 mm^2)

The letter "o" on my typewriter has an area of about 1.8 square millimetres.

STEP 8

0.000 000 1 square metres (10^{-7} m^2)
100 000 square micrometres (10^5 μm^2)

The period on my typewriter takes up an area of about 0.2 square millimetres, or about 200 000 square micrometres.

At this point, I will stop the downward progression. We are approaching the limit of the visible and, past that, there is little new that we can point to, beyond what we have encountered in the downward steps of the ladder of length.

Still, by adding 7 steps to what we have already dealt with in moving upward from the square metre, we have, in a total of 33 steps covering 33 orders of magnitude, gone from the area of the dot at the end of a sentence to the surface area of the largest red giant.

The Ladder of Volume

UPWARD

UPWARD

STEP 1

1 cubic metre (10^0 m^3)

A volume is determined by multiplying a length by a length by a length.

Suppose, for instance, you imagine a cube in which each edge is just 1 metre long. The volume (that is, the space it contains) is 1 metre × 1 metre × 1 metre, or 1 cubic metre, which can be symbolized as 1 m^3.

But what if we had a cube in which each edge was 2 metres long? The volume of this cube is 2 metres × 2 metres × 2 metres, or 8 cubic metres. Thus, when the length of the edge is increased twofold, the volume is increased eightfold. By the same reasoning, if each edge of the cube was 10 metres, the volume would be 1 000 cubic metres. In general, the volume of a series of similar objects is said to be proportional to the cube of the length of a particular edge.

The cubic metre is the basic unit of volume in the SI version of the metric system, and here again round numbers are the rule. Thus, 1 cubic metre is equal to 1 000 cubic decimetres; 1 cubic kilometre is equal to 1 000 000 000 cubic metres; and so on.

In the American system of measurement, we would measure the edge of a cube in inches, feet, or yards, and therefore speak of "cubic inches," "cubic feet," or "cubic yards."

Since 1 foot is equal to 12 inches, 1 cubic foot is equal to 12 × 12 × 12, or 1 728 cubic inches. Since 1 yard is equal to 3 feet, or to 36 inches, 1 cubic yard is equal to 3 × 3 × 3, or 27 cubic feet; and also, to 36 × 36 × 36, or 46 656 cubic inches. Again, the number of Americans who keep such numbers in their heads, or who even know how to work out the numbers if they should need them, are few indeed.

A metre is equal to 1.094 yards; therefore, 1 cubic metre is equal to 1.094 × 1.094 × 1.094, or just about 1.31, or 1 1/3, cubic yards. In reverse, a cubic yard is equal to about 3/4 of a cubic metre. (In addition, a cubic metre is equal to about 35 1/3 cubic feet, and to about 61 000 cubic inches.)

In addition to such measures of volume as cubic inches, cubic feet, and cubic yards, there are other measures that are even more familiar to ordinary Americans—or so they think.

There are "bushels," for instance, each of which holds 4 "pecks," with each peck holding 8 "dry quarts." These measures (and others) are used to describe the volumes of dry materials such as grain. A different set of measures is used for liquids such as water or milk. There we have "gallons," for instance, each of which hold 4 "liquid quarts," each of which holds 2 "liquid pints," each of which holds 4 "gills."

A dry quart is not the same as a liquid quart, mind you. A dry quart is equal to about 1 1/6 liquid quarts. There are innumerable other common volume measures in the United States, even such things as "cups," "tablespoons," and "teaspoons" in cooking. It is very doubtful if even one American out of a thousand has all these measures and their interrelationships straight.

The most familiar of the common units of measure of volume is the liquid quart (which the great majority of Americans refer to simply as "quart," since they have never heard of the dry quart, or have forgotten it if they have). We all know what a quart of milk is, for instance. We can therefore get a notion of the size of the cubic metre if we remember that a cubic metre will hold a little over a thousand (1 057, to be accurate) liquid quarts.

The cubic metre is sometimes called a "stere" (pronounced STEER), from a Greek word meaning "solid." The use of this term, however, is not good form in the SI version of the metric system.

It is not only cubes that have volumes, of course. Spheres, or objects of *any* shape, regular or irregular, have volumes, too, and these can always be expressed in cubic metres.

Suppose, for instance, you have a sphere in which a straight line is drawn from its center to any point on its surface, and this line is found to be 1 metre long. The sphere is then said to have a radius of 1 metre.

To work out the volume of a sphere, we use the formula $4\pi r^3/3$, where "r" is the length of the radius, and π is the usual 3.1416, as an approximate value. The value of $4\pi/3$ is just about 4.189, so we can say that the volume of a sphere is, to a good approximation, $4.189r^3$. Thus, a sphere with a radius of 1 metre has a volume of about 4.189 cubic metres.

(The diameter of a sphere is twice the radius and you can get the correct volume if you set it equal to about $0.523d^3$, where "d" is the length of the diameter.)

If you are dealing with a sphere with a radius of 2 metres, then the volume is $4.189 \times 2 \times 2 \times 2$, or 33.5 cubic metres. As you might expect, doubling the length of the radius (a distance) increases the volume eightfold. Increasing the radius by 10 would increase the volume by 1 000. As for all solids, the volume increases as the cube of the measurement of some appropriate length.

If we ascend the ladder of volume, we will encounter things we came across as we ascended the ladder of length. If, however, we picture volumes always as hollow spheres, with centers located at the center of some astronomical body, we will get a picture different in some respects from that we obtained ascending the ladder of length.

Because of the similarity, however, we can afford to take larger steps in the case of the ladder of volume. We will picture a series of spheres in which the radius increases by an order of magnitude (a factor of ten) at each step. The volume of the sphere will therefore increase by three orders of magnitude (a factor of a thousand).

Thus at Step 1, with which we are beginning, we will imagine a hollow sphere with a volume of 1 cubic metre. The radius of such a sphere is just about 0.62035 metres, or nearly 5/8 metre. The diameter of the sphere would be twice that, or just about 1 1/4 metres. This figure will increase by a factor of 10 with each step upward on the ladder of volume, but we will not refer to it. We will deal only with the volume.

A hollow sphere at Step 1, with a volume of 1 cubic metre, would hold any of innumerable meteoroids which have a longest diameter of 1 1/4 metres or less, if we imagine the center of the sphere at the center of the meteoroid.

Larger meteoroids in this range can survive passage through the atmosphere and will hit the Earth's surface. They can do considerable damage in the immediate vicinity of the collision, damage that could be very serious if it struck in the midst of a population center. However, the Earth offers a large target and, even now, when the Earth is so heavily populated and is so thickly strewn with cities and with other works of humanity, the chances of a damaging strike by a meteorite of considerable size is very small.

The largest meteorite that is actually on view in a museum is in the Hayden Planetarium in New York City. It was found in 1897 on the northwest coast of Greenland by the American explorer, Robert E. Peary (who, twelve years later, was to be the first to reach the North Pole). It is a bit above 4 cubic metres in volume.

One meteorite which is larger, and is, in fact, the largest known, is still buried in northeastern Namibia (in southwestern Africa). Its longest diameter is not quite 3 metres, and it may not be more than 10 cubic metres in volume.

STEP 2

1 000 cubic metres (10^3 m^3)
1 cubic dekametre (10^0 dam^3)

In central Arizona, near the town of Windsor, there is a crater named "Meteor Crater," which was produced by the impact of a meteorite some time between 15 000 and 40 000 years ago. The crater is 1.2 kilometres across and 0.18 kilometres deep. Had the meteorite that had dug that crater in one flashing moment of collision struck even the largest modern city, it would have wiped it out at once. That meteor has been estimated to be about 25 metres in diameter, and it may have been 8 cubic dekametres in volume.

STEP 3

1 000 000 cubic metres (10^6 m^3)
1 cubic hectometre (10^0 hm^3)

Meteor Crater is by no means the largest crater to have made a still-visible mark on the Earth's surface. All the larger craters, however, are much older, and less prominent. They can usually be observed from the air as more or less circular intrusions on the surrounding background, partly water-filled, with their rocky strata bent and distorted, and with differences in vegetation within and without. Many of these must have been made by meteoroids of volume in the Step-3 range.

STEP 4

1 000 000 000 cubic metres (10^9 m^3)
1 cubic kilometre (10^0 km^3)

The largest marking on Earth that we can still make out, and that may have had its origin in a meteorite strike, is located in northern Quebec. The crater is about 440 kilometres across and might have been made by a meteorite with a volume of a cubic kilometre or so.

All the airless objects in the Solar system seem to be littered with craters that arose from collisions with bits of matter, many of them undoubtedly in the cubic-kilometre range, or even far beyond. The only exceptions among the airless worlds are those in which freezing water (as on Callisto), or solidifying lava (as on Io), have filled in the craters.

A sphere in the Step-4 range would be about 1.24 kilometres in diameter and we now can deal with named objects in space. The asteroid Hermes, which was spied uncomfortably close to Earth in 1937, would probably fit into such a sphere.

STEP 5

1 000 000 000 000 cubic metres (10^{12} m^3)
1 000 cubic kilometres (10^3 m^3)

A sphere this size is about 12.4 kilometres across, and would hold an object large enough to be thought of as an asteroid, rather than as a meteoroid. The asteroid Eros, largest of the Earth-grazers, would not quite fit into a Step-5 sphere, since its longest diameter is about 15 kilometres. Its total volume, however, is only about 500 cubic kilometres.

The notion has recently been advanced that the collision with the Earth of an asteroid of about Step-5 volume, some 65 million years ago, cast up such a pall of dust into the stratosphere as to cut off the sunlight long enough to all but sterilize the Earth. It was this event which may have killed off the dinosaurs.

The smallest satellites would fit into a Step-5 sphere. Jupiter's satellite Leda is probably no more than 10 kilometres across its longest diameter and would fit into such a sphere.

Of Mars's two satellites, the smaller, Deimos, is 16 kilometres across its longest diameter and would not fit. However, it is thinner in other directions and its volume is not quite 800 cubic kilometres.

A Step-5 sphere can almost hold the average neutron star, which would have a volume of 1 500 cubic kilometres or so. (Such a star would contain enormously more matter than an asteroid of the same size, of course.)

STEP 6

1 000 000 000 000 000 cubic metres (10^{15} m^3)
1 000 000 cubic kilometres (10^6 m^3)

Up to this point we have been dealing with astronomical objects of irregular shape. By the time we have reached an object with a volume of 1 000 000 cubic kilometres or so, however, the gravitational field is intense enough to force all parts of the object to fall as far toward the center as possible. This results in a sphere.

A Step-6 sphere has a diameter of about 124 kilometres. Here and beyond, a more or less spherical shape is the rule, except for distortions caused by rapid rotation or the nearness of another large object.

Of Jupiter's outermost satellites, Elara, the second largest, is about 80 kilometres across and has a volume of about 270 000 cubic kilometres. The largest, Himalia, with a diameter of 170 kilometres, has a volume of about 2 500 000 cubic kilometres.

STEP 7

1 000 000 000 000 000 000 cubic metres (10^{18} m^3)
1 cubic megametre (10^0 Mm^3)

At this stage, we have a hollow sphere that can hold even the largest asteroids. The Step-7 sphere has a diameter of 1.24 megametres, while the largest asteroid has a diameter of just over 1 megametre and a volume of only a little over 0.5 cubic megametres.

There are satellites that are larger than any asteroid, of course. Saturn's satellite Iapetus has a diameter of 1.44 megametres and a volume of about 1.5 cubic megametres.

STEP 8

1 000 000 000 000 000 000 000 cubic metres (10^{21} m^3)
1 000 cubic megametres (10^3 Mm^3)

With the Step-8 sphere, which has a diameter of about 12.4 megametres, we have passed beyond the satellite stage and reached planetary volumes. Venus has a diameter of 12.1 megametres, and Earth, one of 12.77 megametres. This means that Venus's volume is 926 cubic megametres and Earth's is 1 090 cubic megametres. The two planets neatly straddle the Step-8 sphere.

The sphere will even hold a white dwarf star. The longest-known and most famous white dwarf is Sirius B, the companion of Sirius, the brightest star in the sky. Its volume is about 715 cubic megametres.

STEP 9

1 000 000 000 000 000 000 000 000 cubic metres (10^{24} m^3)
1 000 000 cubic megametres (10^6 Mm^3)

We are now into the realm of the giant planets. Saturn has a diameter of 120 megametres, and Jupiter, one of 143.8 megametres, as compared with a 124-megametre diameter for the Step-9 sphere.

In both planetary cases, however, the diameter gives us a slightly exaggerated notion of the volume. Both Saturn and Jupiter, because of their unusually rapid rotations, have an "equatorial bulge," so that the equatorial diameter, which is the figure usually given, is considerably greater than other diameters. If Saturn had a diameter of 120 megametres in every direction, its volume would be about 916 000 cubic megametres. As it is, it is only about 860 000 cubic megametres. Even so, its volume is seven eighths that of a Step-9 sphere.

As for Jupiter, its volume is about 1 400 000 cubic megametres.

We can even begin to talk about satellite-systems at this stage. If we imagined a Step-9 sphere with its center at the center of Mars, both Martian satellites would be revolving within the sphere. Deimos, the farther satellite, would be revolving at a distance only a little more than a third of the way from the center of the sphere to its surface. A sphere only 1/20 the volume of the Step-9 sphere would, in fact, suffice to hold the Martian satellite-system.

(This does not mean, however, that we can retreat to a Step-8 sphere for the Martian satellites. Since we are taking steps three orders of magnitude at a time, the Step-8 sphere is only 1/1 000 the volume of the Step-9 sphere, or only 1/50 of what would be needed to enclose the Martian satellites.)

STEP 10

1 000 000 000 000 000 000 000 000 000 cubic metres (10^{27} m^3)
1 cubic gigametre (10^0 Gm^3)

Now, after another thousandfold expansion, we reach the cubic gigametre stage and find that our sphere can hold Saturn and its entire ring system without trouble. The extreme diameter of Saturn's outermost "E ring" (not visible from Earth) is not quite 1 gigametre, while the diameter of a Step-10 sphere is 1.24 gigametres.

The Step-10 sphere can also hold Uranus and its entire satellite system, since the orbit of Oberon, Uranus's outermost known satellite, has a diameter of only 1.17 gigametres. For that matter, it can hold the Earth-Moon system, since the orbit of the Moon around the Earth has a mean diameter of 0.76 gigametres.

Only one object in the Solar system is large enough, in itself, to be in this range, and that, of course, is the Sun. It has a diameter of about 1.4 gigametres, and its volume is therefore about 1.44 cubic gigametres.

STEP 11

1 000 000 000 000 000 000 000 000 000 000 cubic metres (10^{30} m^3)
1 000 cubic gigametres (10^3 Gm^3)

A Step-11 sphere can almost hold the Neptunian satellite system. Neptune's major satellite, Triton, has an orbit that would fit comfortably into a Step-10 sphere, but the outer satellite, Nereid, has an orbit with a diameter of 11.1 gigametres. This could easily fit into the 12.4-gigametre diameter of the Step-11 sphere, if Nereid had an essentially circular orbit, as Triton has. However, Nereid's orbit is quite eccentric, and at its farthest recession from Neptune, it would move no less than 3.5 gigametres beyond the surface of the sphere.

The Step-11 sphere could also hold the entire Saturnian satellite system except for the outermost satellite, Phoebe. Leaving Phoebe out of account, the extreme diameter of the Saturnian satellite system is only 7.1 gigametres.

A Step-11 sphere will not hold even the relatively small red-giant star, Arcturus, which, with a diameter of 37 gigametres, has a volume of more than 25 000 cubic gigametres.

STEP 12

1 000 000 000 000 000 000 000 000 000 000 000 cubic metres (10^{33} m^3)
1 000 000 cubic gigametres (10^6 Gm^3)

At the million-cubic-gigametre stage, every satellite system can be contained. Jupiter has the largest satellite system of any planet, but even its outermost satellite has an orbital diameter of no more than 48 gigametres as compared to the 125-gigametre diameter of the Step-12 sphere. A sphere with only 1/17 the volume of the Step-12 sphere would suffice to hold the Jovian satellite system.

In fact, we edge into the planetary system now. If the Step-12 sphere had its center at the center of the Sun, it would contain the orbit of Mercury, the nearest planet, if that orbit were circular. The diameter of Mercury's orbit is about 115 gigametres, but the eccentricity of that orbit would place Mercury up to 7.5 gigametres beyond the surface of the Step-12 sphere at one end.

The red giant, Beta Pegasi, with a diameter of 150 gigametres, will not quite fit into a Step-12 sphere.

STEP 13

1 000 000 000 000 000 000 000 000 000 000 000 000 cubic metres (10^{36} m^3)

1 cubic terametre (10^0 Tm^3)

If the center of a Step-13 sphere were placed at the center of the Sun, its surface would extend almost to the orbit of Jupiter. Jupiter's orbit has a diameter of 1.56 terametres, so that a sphere that would just include that orbit would have a volume of about 2 cubic terametres.

The large red giant, Betelgeuse, at its pulsating largest, would just fit within the sphere.

STEP 14

1 000 000 000 000 000 000 000 000 000 000 000 000 000 cubic metres (10^{39} m^3)

1 000 cubic terametres (10^3 Tm^3)

This expansion carries the diameter to 12.4 terametres, and the Step-14 sphere, if centered on the Sun's center, is large enough to include the entire planetary system, since the diameter of the orbit of Pluto, the farthest known planet, is only 11.8 terametres.

Pluto's orbit, however, is even more eccentric than is that of Mercury. At one end of its orbit, it recedes as far as 7.22 terametres from the Sun, and, at that point, it is very nearly a full terametre beyond the surface of the Step-14 sphere.

If we consider individual stars, even the largest red giant would not fill more than 1/90 of this sphere.

STEP 15

1 000 000 000 000 000 000 000 000 000 000 000 000 000 000 cubic metres (10^{42} m^3)
1 000 000 cubic terametres (10^6 Tm^3)

The Step-15 sphere, if centered on the Sun's center, will include the orbits of all but one of the comets whose orbits are known. The exception, of course, is Comet Kohoutek.

If we imagine the Step-15 sphere to be centered on the Sun's center, Comet Kohoutek will, at that end of its orbit which is most distant from the Sun, retreat far beyond the surface of the sphere. In fact, the sphere would include less than 1/9 the length of the enormously elongated orbit of Comet Kohoutek.

Even if we placed the center of the Step-15 sphere (124 terametres in diameter) at the center of Comet Kohoutek's orbit, that orbit would extend 200 terametres beyond either side of the surface of the sphere. More than two thirds of its orbit would lie beyond the sphere's surface.

STEP 16

1 000 000 000 000 000 000 000 000 000 000 000 000 000 000 000 cubic metres (10^{45} m^3)
1 cubic petametre (10^0 Pm)

At the Step-16 volume, we reach the cubic-petametre stage, and now the entire mighty orbit of Comet Kohoutek falls within the sphere, if the latter is centered on the Sun. Comet Kohoutek, at the far end of its orbit, is 0.5282 petametres from the Sun, while the sphere, with a diameter of 1.24 petametres, has its surface 0.62 petametres from the Sun in every direction.

STEP 17

1 000 000 000 000 000 000 000 000 000 000 000 000 000 000 000 000 cubic metres (10^{48} m^3)
1 000 cubic petametres (10^3 Pm^3)

The Step-17 sphere has a diameter of 12.4 petametres, which is equal to 1.31 light-years. The volume of the sphere is very nearly 1.18 cubic light-years, therefore. We won't be too far off if we say that 1 000 cubic petametres is equal to 1 cubic light-year.

If the Step-17 sphere were centered on the Sun, the surface of the sphere would be located along the innermost reaches of the great shell of comets that may be circling the Sun, according to Oort's theory.

STEP 18

1 000 000 000 000 000 000 000 000 000 000 000 000 000 000 000 000 000 cubic metres (10^{51} m^3)
1 000 000 cubic petametres (10^6 Pm^3)

The Step-18 sphere, with its mighty diameter of 13.1 light-years, would, if centered on the Sun, stretch its surface far beyond the Sun's shell of comets. It would, in fact, include the 30 stars nearest the Sun within its thousand-cubic-light-year volume.

If the sphere were centered at the center of a small globular cluster in the constellation Sagitta, it would include almost the entire cluster, or about 15 000 stars, which would be spread 500 times as thickly through space as stars are in our part of the Universe.

This means that the average distance between stars in such a globular cluster would be only about an eighth of what it is here. In our neighborhood, the nearest star to the Sun is 4.4 light-years away; in a globular cluster, it might be only 0.5 light-years away. Even at this distance (small for a star, by our standards), stars would look like mere points of light. The stars in the sky within a globular cluster would be much brighter than they are here, but they would seem no larger.

STEP 19

1 000 000 000 000 000 000 000 000 000 000 000 000 000 000 000 000 000 000 cubic metres (10^{54} m^3)
1 cubic exametre (10^0 Em^3)

A Step-19 sphere (with a diameter of 131 light-years), placed at the center of even the largest known globular cluster, would easily enclose it all. Several million stars would be included within it.

STEP 20

1 000 000 000 000 000 000 000 000 000 000 000 000 000 000 000 000 000 000 000 cubic metres (10^{57} m^3)
1 000 cubic exametres (10^3 Em^3)

Suppose we place the Step-20 sphere, with its 1 310-light-year diameter, at the center of our Galaxy. It would then be at the center of what is essentially the equivalent of a gigantic globular cluster. The sphere would contain several billions of stars.

STEP 21

1 000 000 000 000 000 000 000 000 000 000 000 000 000 000 000 000 000 000 000 000 cubic metres (10^{60} m^3)
1 000 000 cubic exametres (10^6 Em^3)

A Step-21 sphere is large enough to enclose a dwarf galaxy. If its center is placed at the center of the Small Magellanic Cloud, its 13 100-light-year diameter would include that entire galaxy.

Of course, the Small Magellanic Cloud is a dwarf galaxy, and it has relatively few stars which its overall gravitational field does not succeed in squeezing into a relatively small volume. The same sphere, with its center at the center of our own Galaxy, would enclose over 100 billion stars.

STEP 22

1 000 000 000 000 000 000 000 000 000 000 000 000 000 000 000 000 000 000 000 000 000 cubic metres (10^{63} m^3)
1 000 000 000 cubic exametres (10^9 Em^3)

At this stage, we ought to be using a new prefix beyond the "exa-," but the SI version does not yet supply one. We must therefore continue using exametres.

A Step-22 sphere with its center at the center of our Galaxy would spread out its vast 131 000-light-year diameter to include the entire swarm of up to 300 000 000 000 stars.

Yet even though our Galaxy is a giant among galaxies, it is by no means the largest. There are enormous galaxies that contain up to 10 trillion stars, with diameters three times the Step-22 sphere.

STEP 23

1 000 000 000 000 000 000 000 000 000 000 000 000 000 000 000 000 000 000 000 000 000 000 cubic metres (10^{66} m^3)
1 000 000 000 000 cubic exametres (10^{12} Em^3)

Any galaxy, even the largest, would be lost within a Step-23 sphere. Such a sphere, 1 310 000 light-years in diameter, if centered at the center of our Galaxy, would include not only our Galaxy but about five of the nearby dwarf galaxies.

STEP 24

1 000 000 000 000 000 000 000 000 000 000 000 000 000 000 000 000 000 000 000 000 000 000 000 cubic metres (10^{69} m^3)
1 000 000 000 000 000 cubic exametres (10^{15} Em^3)

A Step-24 sphere with its center at the center of our Galaxy would hold the entire Local Group within itself, and with a great deal of room to spare. Indeed, even the largest clusters of galaxies, containing a thousand galaxies or so, would fit within such a sphere.

STEP 25

1 000 000 000 000 000 000 000 000 000 000 000 000 000 000 000 000 000 000 000 000 000 000 000 000 cubic metres (10^{72} m^3)
1 000 000 000 000 000 000 cubic exametres (10^{18} Em^3)

A Step-25 sphere with its center in our Galaxy (anywhere in our Galaxy, for the exact place would no longer make a perceptible difference) would include within it over 20 000 galaxies.

STEP 26

1 000 000 000 000 000 000 000 000 000 000 000 000 000 000 000 000 000 000 000 000 000 000 000 000 000 cubic metres (10^{75} m^3)
1 000 000 000 000 000 000 000 cubic exametres (10^{21} Em^3)

A Step-26 sphere with its center in our Galaxy would include within itself over 20 000 000 galaxies.

STEP 27

1 000 000 000 000 000 000 000 000 000 000 000 000 000 000 000 000 000 000 000 000 000 000 000 000 000 000 cubic metres (10^{78} m^3)
1 000 000 000 000 000 000 000 000 cubic exametres (10^{24} Em^3)

The Step-27 sphere would have a diameter of 13 100 000 000 light-years, and if its center were in our Galaxy, it would include within itself perhaps 20 000 000 000 galaxies.

And with that, we run out of Universe. As nearly as we can tell, the observable Universe is twice as wide as a Step-27 sphere and has eight times the volume and would contain perhaps 100 000 000 000 galaxies.

If we were to go on to a Step-28 sphere, however, it would be over 5 times the diameter of the Universe and over 150 times the volume. Thus, in 26 steps, covering 78 orders of magnitude, we have gone from the volume of a small meteorite to the volume of the Universe.

The Ladder of Volume

DOWNWARD

DOWNWARD

STEP 1

1 cubic metre (10^0 m^3)

Let us now go back to 1 cubic metre, and before beginning a brief descent, let us consider it in relation to the human body.

Imagine a hollow box about 1 3/4 metres high, 1/2 metre wide, and 1 1/6 metres deep. This would have a volume of about 1 cubic metre. I could step into such a box and find it just about high enough and wide enough to hold me. I could step forward, and three more people just like myself could step in behind me. In short, such a box would hold four people like myself.

STEP 2

0.001 cubic metres (10^{-3} m^3)
1 cubic decimetre (10^0 dm^3)

A unit of volume which is large enough to hold four human bodies is, perhaps, too large to be entirely convenient as a base upon which to build other units of the sort. There has been considerable pressure, therefore, for a smaller base measure of volume.

The SI version might feel justified in holding rigidly to the cubic metre, since that ties in tightly with the square metre as the basic unit of area, and the metre as the basic unit of length. The pressure, however, has been irresistable, and the cubic decimetre, which is 1/1 000 the volume of the cubic metre, has been given special consideration.

A cubic decimetre is commonly called a "litre" from the name of an old French unit of volume, and this, pronounced "LEE-ter" and symbolized as "L," can also be used as a basic unit of volume.

As it happens, 1 litre is equal to 1.057 liquid quarts, and, in reverse, a liquid quart is equal to 0.9464 litres. Thus, a litre (or a cubic decimetre) is approximately equal to a liquid quart. If you visualize the common quart container of milk (easy to do in the United States), you are thinking of something which represents just slightly less than a litre, or cubic decimetre, of volume. What's more, a gallon is, roughly speaking, four litres, and a pint is, just as roughly speaking, half a litre.

STEP 3

0.000 001 cubic metres (10^{-6} m^3)
1 cubic centimetre (10^0 cm^3)

The smallest unit of volume based on a measure of length in the ordinary American system is the cubic inch. Since an inch is equal to 2.54 centimetres, a cubic inch is equal to 2.54 centimetres × 2.54 centimetres × 2.54 centimetres, or 16.4 cubic centimetres. In reverse, a cubic centimetre is equal to 1/16.4 cubic inches.

A liquid quart is divided into 32 "fluid ounces." Since a quart is equal to 946.4 cubic centimetres, a fluid ounce is equal to 1/32 of that, or 29.6 cubic centimetres.

Druggists (for whom "apothecaries" is an old-fashioned term) have long been accustomed to deal with very small volumes of powerful drugs, so there is something called "apothecaries fluid measure" to deal with such small volumes. It has long been displaced by the metric system, but its units are still part of the English language.

Thus, a fluid ounce is divided into 8 "fluid drams," and each fluid dram is equal to 3.7 cubic centimetres. The fluid dram, in its turn, is divided into 60 "minims," and each minim is therefore about 1/16 of a cubic centimetre. The minim is usually taken to represent a drop of water, so that there would be about 16 drops of water to a cubic centimetre.

Since a cubic centimetre is a thousandth of a cubic decimetre, or of a litre, it is sometimes referred to as a "millilitre."

STEP 4

0.000 000 001 cubic metres (10^{-9} m^3)
1 cubic millimetre (10^0 mm^3)

Since a cubic millimetre is a thousandth of a cubic centimetre, it follows that a drop of water represents about 60 cubic millimetres.

With that, we will stop the descent. Going further will not bring us anything very new in comparison with what we experienced in descending the ladder of length, and it is now time to change the subject considerably.

The Ladder of Mass
UPWARD

UPWARD

STEP 1

1 kilogram (10^0 kg)

So far, we have dealt with length, with area (the square of length), and with volume (the cube of length). Now let us switch to another kind of measurement that has nothing to do with length and that is just as fundamental as length. That is "mass."

Nonscientists usually think of mass as "weight," and as long as we confine ourselves to the sea-level surface of the Earth, the two, mass and weight, are indeed directly proportional to each other to a considerable degree of precision. That is, if object A has 4.23 times as much mass (or "is 4.23 times as massive") as object B, then object A also has 4.23 times as much weight (or "is 4.23 times as heavy") as object B.

This proportion holds so well over all ranges that even scientists interchange the terms on occasion, and tend to use the same set of units for both.

Nevertheless, mass and weight are completely different. Mass is the measure of the ease with which an object can be accelerated. (To put it more simply, but not *quite* accurately, mass is the measure of the amount of matter contained by an object.) Weight, on the other hand, is the measure of the force with which the gravitational field of a nearby body (under ordinary conditions, almost invariably the Earth) attracts an object.

As it happens, the greater the mass of a body, the greater the force of gravitational attraction by the Earth upon that body. This is a general rule, so that we might think that the greater the mass of a body, the greater the weight—inevitably. In that case, the two would be identical.

But that isn't so. The force of gravitational attraction by the Earth upon a particular body also depends upon the distance of that body from the center of the Earth, while the mass does not. Therefore, if an object is taken from a valley to a mountaintop, the weight decreases slightly, while the mass does not.

Again, if an object is taken from the Earth to the Moon (which has a surface gravity only one sixth that of the Earth), the weight decreases to one sixth its former amount, but the mass remains unchanged. If an object is in free fall, as it would be if it were an orbiting satellite or any of its contents, its weight would be essentially zero, but, again, its mass would be what it always was.

For that reason, scientists are careful to distinguish between mass and weight, and to use mass only when they mean mass, and weight only when they mean weight. Since mass is the more constant of the two, they tend to use mass far more often than weight.

(Mass is not absolutely changeless. It changes with motion, but in order for the change to be noticeable, motion must be extreme—very fast. Nothing human-made moves quickly enough to produce a noticeable effect on mass.)

When the founders of the metric system, in the 1790s, searched for a unit of mass, they made an effort to tie it in with the units of length according to the system they had already set up.

If we begin with the centimetre, a common unit of length in the metric system, we can build on it the cubic centimetre as a unit of volume. Imagine a cubic centimetre of pure water under certain specified environmental conditions we need not go into here. The mass of that quantity of water was set equal to 1 "gram." (The word, gram, is from a Greek word meaning a letter of the alphabet, presumably because a Greek weight had a Greek letter on it—the letters also served as numbers to the Greeks—to indicate the size of that weight.) The gram is symbolized as "g."

The gram was taken as the basic unit of mass, and to it were added the usual prefixes. A hectogram is a hundred grams, a milligram is a thousandth of a gram, and so on.

The gram, as it happens, is a rather small unit of mass. One gram is equal to 0.002205 of the "pounds" that are commonly used as units of mass in the American system. One gram is equal to 1/453.5 pounds, therefore.

(Notice that pounds are usually thought of as units of weight and are so used. So are grams, for that matter. Scientists are stuck with that identity of units, even though the proper units of weight are units of force, something we won't go into. When scientists are forced to use the units of mass as units of weight, they sometimes distinguish the two by speaking of so many "pounds [w]," or of so many "grams [w]." In this section of the book, we will be dealing with mass only, and never with weight, so we won't have any problem.)

The gram is, in fact, too small for ordinary use. We in the United States tend to think in pounds because these measure quantities of matter that we are likely to encounter in ordinary life. If we think in terms of the metric system, we would ordinarily encounter materials in quantities of thousands of grams.

The "kilogram" is equal to 1 000 grams, and that is a much more convenient unit to use as the basic unit of mass. It is so used in the SI version of the metric system. Since a cubic decimetre (or litre) is equal to 1 000 cubic centimetres, it follows that 1 litre of water has a mass of 1 kilogram.

A kilogram is equal to 2.2046 pounds or, approximately, 2 1/5 pounds.

To indicate the handiness of a kilogram, a small frying chicken, as sold by the butcher, might weigh 1 kilogram. A quart of milk weighs very nearly 1 kilogram. The lungs of a human adult have an average mass of about 1 kilogram, while the adult human brain has an average mass of about 1.45 kilograms.

Beginning with 1 kilogram, then, we will move up the ladder to larger and larger masses. Since mass is as fundamental a type of measure as length is, I will as in the case of length move up half an order of a magnitude at a time.

STEP 2

3.16 kilograms ($10^{0.5}$ kg)

The average mass of a newborn male infant in the United States is about 3.4 kilograms; of a female infant, about 3.36 kilograms. A gallon of water has a mass of 3.785 kilograms, and the total blood content of a human being of average size is about 5 kilograms.

At this mass-level, we can even deal with the Earth's atmosphere, after a fashion. The atmosphere consists of a mixture of a number of gases, of which the chief are nitrogen and oxygen. There are a number of minor constituents, including the six so-called noble gases, which do not ordinarily react with other substances. Of these, the rarest is radon, which is radioactive so that the atoms break down at a fixed rate and cease being radon. Of any quantity of radon, half the atoms break down in just under four days. The only reason that radon exists at all is that more is continually formed as the element uranium breaks down very slowly. A balance is maintained between the formation of radon and its breakdown, so that about one part in ten million trillion of any given volume of air is radon. This means that all the radon contained in the atmosphere of the Earth as a whole comes to about 2.5 kilograms, less than the mass of a newborn baby.

STEP 3

10 kilograms (10^1 kg)

There is a wide variety of units of mass (or weight) in the world, dating back to the time when every region, nation, and even community set its own units and standards. Few of them have survived into the modern era of the metric system, but, as one might expect, there are vestigial remnants, particularly in Great Britain and the United States, where resistance to the metric system was most long-lasting.

In various places, there were units of mass called "stone" (or the equivalent in different languages). Clearly, this name originated from the fact that some particular stone would be used as the standard against which all masses were measured. The stone varied in quantity from region to region, and one that remains in use today is the British stone, which is commonly used in Great Britain in giving the mass of human beings. Thus, I could be spoken of as having a 12-stone mass.

The British stone is 14 pounds, or 6.36 kilograms, which is on the lower side of the Step-3 level of mass.

One land animal, and only one, has a brain more massive than that of the human. That is the elephant, and the largest elephant brain ever measured had a mass of about 7.5 kilograms, five times that of the human brain, and well above the average for elephants generally.

Still more massive are the brains of some of the large whales. The largest brains ever measured (and probably that ever existed) are those of sperm whales. One such brain had a mass of 9.2 kilograms, about 6 1/3 times the mass of a human brain, and almost as massive as an adult human skeleton, which comes to about 10 kilograms.

Adult tomcats have masses in the 10-kilogram range. So does a year-old human baby.

At the Step-3 mass-range, we are approaching the limit of flying birds. The problem of flight, of supporting a mass on thin air by the rapid movement of wings, is a difficult one, and becomes rapidly more difficult as mass increases. For that reason, flying animals are far less massive on the whole than swimming or walking animals. The largest and best fliers in the animal kingdom are the birds, but even the largest flying bird is not much more massive than 15 kilograms.

The largest eggs produced by any creature, living or extinct, are those of the Aepyornis (or elephant bird) that once lived in Madagascar and which I described earlier. Its eggs were up to 12 kilograms in mass, seven times the mass of ostrich eggs. The largest seeds are the double coconuts of the coco-de-mer tree of the Seychelle Islands, which have a mass of up to 18 kilograms.

STEP 4

31.6 kilograms ($10^{1.5}$ kg)

The average mass of an American ten-year-old child is at this level. The largest monkey is the mandrill, and adult male mandrills have an average mass of about 30 kilograms. Some of the well-known predators are in this mass-range; the ocelot is an example.

It is possible that the extinct flying reptiles, the pterosaurs, may have, in some cases, been more massive than the most massive flying birds of today. Some pterosaurs may have had a mass of 25 kilograms or so, but there is some question as to whether they were true flyers. They may simply have been gliding animals, taking advantage of updrafts to gain altitude.

STEP 5

100 kilograms (10^2 kg)

Here we reach the upper limit of the mass of normal human beings. The average adult American female is about 58 kilograms in mass (my wife is 55.5 kilograms, though she is taller than average). The average adult male is about 68 kilograms in mass (and I am 77 kilograms). Males at the upper end of the height range, if they are solidly built, may reach 100 kilograms without being unduly fat.

For the most part, however, people with masses of more than 100 kilograms are decidedly fat, if not obese. The famous fictional detective, Nero Wolfe, was frequently described by the narrator, Archie Goodwin, as being a seventh of a ton in weight. This is equivalent to a mass of 130 kilograms.

Large dogs are as massive as large men. The most massive breed of dog is the Saint Bernard, and one of these has been reported as 140 kilograms in mass.

Mountain lions have masses of up to 130 kilograms. The largest living bird, the ostrich, can have a mass of up to 150 kilograms, while the largest living primate, the gorilla, can have a mass of up to 180 kilograms.

Even objects we think of as light have masses that mount up if enough quantity is involved. We think of air as possessing little mass, but the air contained in a good-sized living room in a luxury New York apartment would have a mass in the neighborhood of 100 kilograms.

STEP 6

316 kilograms ($10^{2.5}$ kg)

At the Step-6 range of mass, we are at the absolute human limit. There have been cases of women who have finally reached masses in excess of 360 kilograms, and men who attained masses of over 450 kilograms, but these are likely to end up on display in circuses. It is an enormously abnormal situation.

The most massive of the large cats are the lions and tigers. Lions can attain a mass of up to 313 kilograms, and tigers up to 350 kilograms. In their cases, however, the masses are normal and do not interfere with their efficient functioning.

Although no living bird (flightless, of course) can reach the Step-6 mass-range, there were birds larger than the ostrich that are now extinct but were alive in historic times. The giant moa of New Zealand, which did not become extinct till nearly modern times, attained a mass of 225 kilograms. The Aepyornis of Madagascar, which was perhaps alive until the seventeenth century, and which is the most massive of all known birds, living or extinct, may have reached a mass of as much as 450 kilograms.

The largest known primate, now extinct, was Gigantopithecus, which resembled an outsized gorilla and which may have attained a mass of up to 270 kilograms, four times the mass of an average American male.

STEP 7

1 000 kilograms (10^3 kg)
1 megagram (10^0 Mg) or 1 tonne (10^0 t)

The Step-7 level of mass places us in the range of the megagram, which is equal to a million grams.

We are also in the range of a unit of mass, familiar in the United States, which is called the "ton" and which is equal to 2 000 pounds. This is sometimes called the "short ton" because in Great Britain the comparable unit, often used, is the "long ton," which is equal to 2 240 pounds. The short ton is thus about 0.907 megagrams and the long ton is about 1.016 megagrams.

In the metric system, the megagram is a compromise between the two tons and, in French, 1 megagram, or 1 000 kilograms, is a "tonne." Unfortunately, "tonne" has the same pronunciation as "ton" and to avert the possibility of confusion, the tonne is known as the "metric ton" in the United States.

The SI version of the metric system prefers megagram, but reluctantly permits the use of the metric ton. There is the danger that, in the United States, this will be confused with the common ton, which is ten percent smaller, and for that reason I am tempted to stick to the megagram. However, if one uses the metric ton, one can add prefixes and reach levels six orders higher than the same prefixes would reach if affixed to "gram." That will be none too high as we shall see as we ascend the ladder of mass. Furthermore, I will use "tonne" rather than "metric ton," for it is much more convenient to say "kilotonne" and "megatonne" than "metric kiloton" and "metric megaton." The chance of confusion between "tonne" and "ton" will simply have to be taken. Think metric, and you won't be confused.

The largest land carnivore is the Kodiak bear of Alaska, specimens of which, in the wild, can reach a mass of 1.2 tonnes, while a caged bear of this type, able to gain weight through good feeding and a sedentary life, is reported to have reached a mass of 1.65 tonnes.

The largest aggressively carnivorous fish (one that eats animals of considerable size, as opposed to the tiny animals that are nearly or quite microscopic and are called "plankton") is the great white shark, and that can have a mass of up to 1 tonne.

On the whole, animals that live on smaller animals can grow to greater weights than those that live on larger ones, simply because there are more smaller animals than larger ones. Animals that live on plants can grow larger still.

Thus sturgeon, which live on small animals, can grow even larger than the sharks, and some attain a mass of up to 1.4 tonnes.

Similarly, the large bears can outmass the large cats, for instance, because bears are omnivorous and can supplement a diet of meat with vegetation, which the large cats cannot do.

Among those land animals that are strictly herbivorous, examples of those in the Step-7 range are the moose (the largest of the deer), which can reach a mass of perhaps 0.8 tonnes; the eland (the largest of the antelopes), which has been reported up to 0.94 tonnes; and the giraffe, which can reach a mass of about 1.05 tonnes.

STEP 8

3 160 kilograms ($10^{3.5}$ kg)
3.16 tonnes ($10^{0.5}$ t)

At the Step-8 mass-range, we leave living land carnivores behind and must pass on to the really large herbivores. Both the rhinoceros and the hippopotamus may attain masses as high as 4.5 tonnes. Among sea creatures, the giant squid, which is the largest of all invertebrates, may attain this mass-range as well.

There are human artifacts at every mass-range we have been discussing, but up to this point, there has been nothing so distinctive out of all the crowds of them as to force mention. Here, though, it should be pointed out that even people without the benefit of any but the simplest machines have handled remarkable masses of material. The Great Pyramid, built about 2600 B.C., is composed of 2 300 000 blocks of stone of an individual mass (on the average) of 2.5 tonnes.

STEP 9

10 000 kilograms (10^4 kg)
1 dekatonne (10^0 dat)

The largest land carnivores that ever lived (now extinct, of course) were the fearsome *Tyrannosaurus rex* and allied "carnosaurs." These may have reached a mass of 0.7 dekatonnes, nearly six times that of a Kodiak bear and ten times that of a large tiger.

The only living land animals with masses in the Step-9 mass range are herbivores. There are two species of elephants, the Asiatic and the African. The Asiatic is the smaller of the two, and the males may approach a mass of 0.8 dekatonnes, distinctly larger than any land carnivore known to have lived at any time. The African elephant is larger still, and males may be as massive as 1 dekatonne.

Some extinct species of elephants may have been even more massive, but the most massive mammal that ever lived was not an elephant, but a member of the rhinoceros family. This was the *Baluchitherium,* which may have had a mass of 2 dekatonnes, twice that of the largest elephant.

It is possible for sea animals to grow larger and more massive than land animals do. The buoyancy of the water virtually negates any overall gravitational pull on the animal, so that the problem of lifting the body upward against gravity and moving it, which exists on land and grows rapidly more difficult of solution as mass increases, does not exist in water. (Then, too, food is more plentiful in the sea.)

Thus, the largest mammals are whales that live exclusively in the sea. Even a small species of whale is likely to be as massive as an elephant.

Step 10

31 600 kilograms ($10^{4.5}$ *kg)*
3.16 dekatonnes ($10^{0.5}$ *dat)*

The Pacific gray whale may attain a mass of 4 dekatonnes, while the whale shark (which is not a whale, but a fish, and the largest of all living fish) may also attain such a mass.

The whale shark, though 40 times as massive as the great white shark, is not 40 times as terrifying; in fact, it is not terrifying at all. It lives on plankton, which is the equivalent of saying that a land animal lives on grass. It is a nonaggressive creature that is totally harmless except for what damage it might do by accidental collision.

Some of the larger herbivorous dinosaurs (now all extinct) were also in the Step-10 mass-range. So is the largest meteorite on display, the one in the American Museum of Natural History in New York, for it has a mass of nearly 3.1 dekatonnes.

Stonehenge, in southwestern England, is constructed of large rocks with masses of up to 4.5 dekatonnes. It was built between 1800 and 1400 B.C., and is an example of what human muscles, allied to human ingenuity, can do.

STEP 11

100 000 kilograms (10^{5} *kg)*
1 hectotonne (10^{0} *ht)*

We have now reached the limit of land-animal mass.

The largest land animal ever to have lived was a gigantic dinosaur (now extinct) known as *Brachiosaurus*. It may possibly have reached a mass of 0.8 hectotonnes, according to some estimates. This is four times the mass of a *Baluchitherium*.

Only the large whales can match this, and these are still alive. The largest carnivorous whale (and the largest carnivorous animal that ever lived, in all likelihood) is the sperm whale, where masses of up to 0.72 hectotonnes have been recorded.

The fin whale, which is a plankton-eater, can reach a mass of very nearly 1 hectotonne. The record, however, is held by the blue whale of Antarctic waters, which is thought to approach 1.9 hectotonnes in mass, in some cases. There is no question but that the blue whale is the largest living animal, and, very likely, the largest animal that ever lived, although an extinct shark may possibly rival it.

The largest meteorite we know of, the one in Namibia, has a mass of about 0.6 hectotonnes.

STEP 12

316 000 kilograms ($10^{5.5}$ kg)
3.16 hectotonnes ($10^{0.5}$ ht)

We have passed beyond animal life, but not life as a whole. The plant world can produce specimens that are far more massive than any animal. Any bulky tree that is 30 metres tall would rival the largest whales in mass.

A stone of equivalent size would be more massive still. The ancient Egyptians built monolithic obelisks (tall, tapering structures built of a single rock) which may conceivably have been inspired by tree trunks first used as monuments. The largest obelisk now stands in Rome. It is 33.5 metres tall, and has a mass of 5 hectotonnes. Compare this with New York's famous Statue of Liberty (hollow, to be sure), which weighs only 2 hectotonnes.

Before being too lost in wonder at ancient artifacts, however (much though they deserve it), consider that a giant Boeing Jumbo plane, fully loaded, has a mass of about 3.85 hectotonnes, and, in 1979, was made to *fly*.

STEP 13

1 000 000 kilograms (10^{6} kg)
1 kilotonne (10^{0} kt)

The preferred way of representing a million kilograms in the SI version of the metric system is as "1 gigagram," which can be symbolized as 1 "Gg," but clinging to tonnes will give me a further reach, and I will therefore use the kilotonne at this point.

At Step 13, we are into the range of the very largest trees, and of these the sequoia trees of California hold the pride of place, for they are the most massive trees, though not the tallest. A sequoia tree known as "General Sherman" in Sequoia National Park has been estimated to have a mass of 2 kilotonnes.

STEP 14

3 160 000 kilograms ($10^{6.5}$ kg)
3.16 kilotonnes ($10^{0.5}$ kt)

There are two kinds of meteors that, together, make up all but a couple of percent of those that have been recovered on Earth. The more common is a stony variety, and the less common a nickel-iron variety.

A stony meteor, roughly spherical and about 12.5 metres in diameter, would weigh about 3 kilotonnes. An iron meteor would weigh that much if it were half the diameter. These would be a hundred times the mass of the largest meteorite on display.

There are, of course, a substantial number of such 3-kilotonne objects (and larger ones, too) in space, but their collisions with Earth, while not impossible, are very rare.

STEP 15

10 000 000 kilograms (10^{7} kg)
10 kilotonnes (10^{1} kt)

The Eiffel tower of Paris was built in 1889 and remained the highest structure in the world for 41 years. It has a mass of 8 kilotonnes, four times that of the largest living object.

The noble gas xenon is the rarest of the stable noble gases; that is, of those noble gases that, unlike radon, do not break down. Less than one part per ten billion of any volume of pure, dry air is made up of xenon. Even so, the total mass of xenon in that portion of the atmosphere which rests upon the state of Rhode Island comes to 10 kilotonnes, 1.25 times that of the mass of the Eiffel tower.

STEP 16

31 600 000 kilograms ($10^{7.5}$ kg)
31.6 kilotonnes ($10^{1.5}$ kt)

We are now in the mass-range of modern ocean liners. In 1904, the British built the liner *Baltic,* which continued to ply the seas till 1933. I mention it because, in 1923, it carried among its passengers my father, my mother, my sister, and myself as immigrants to the United States. Its mass was about 21.5 kilotonnes. The *Lusitania,* which was launched in 1907 and which sank as a result of a German torpedo in 1915, had a mass of about 28.5 kilotonnes.

STEP 17

100 000 000 kilograms (10^8 *kg)*
100 kilotonnes (10^2 *kt)*

The most massive of the ocean liners was the *Queen Elizabeth,* launched in Great Britain in 1940 and taken out of service in 1972. It had a mass of about 76 kilotonnes.

Still larger are the largest aircraft carriers. The U.S.S. *Nimitz,* launched in 1972, has a mass of about 83 kilotonnes.

The population of Costa Rica has a total mass approaching 100 kilotonnes, if we assume that the average human being weighs 50 kilograms.

STEP 18

316 000 000 kilograms ($10^{8.5}$ *kg)*
316 kilotonnes ($10^{2.5}$ *kt)*

Oil tankers these days are larger than anything else at sea, and the largest, completed in 1981 by a Japanese firm, has a mass of about 510 kilotonnes. This mass is more than 6 times that of the U.S.S. *Nimitz.*

The population of Switzerland has a total mass of perhaps 316 kilotonnes.

STEP 19

1 000 000 000 kilograms (10^9 *kg)*
1 megatonne (10^0 *Mt)*

At the Step-19 mass-level, we are in the range of very large skyscrapers and very small asteroids. An asteroid, roughly spherical in shape and about 87.5 metres in diameter, would have a mass of 1 megatonne. The population of Rumania would also have a total mass of about 1 megatonne.

STEP 20

3 160 000 000 kilograms ($10^{9.5}$ kg)
3.16 megatonnes ($10^{0.5}$ Mt)

The population of Mexico has a total mass of about 3 megatonnes.
The Great Pyramid of Egypt has a mass estimated to be about 6.3 megatonnes, nearly 800 times the mass of the Eiffel tower, and 12.3 times the mass of the largest floating vessel at sea.

STEP 21

10 000 000 000 kilograms (10^{10} kg)
10 megatonnes (10^{1} Mt)

The population of the United States has a total mass of about 10 megatonnes. The Grand Coulee Dam on the Columbia River in the state of Washington required the use of a mass of 19.5 megatonnes of concrete, which is three times the mass of the Great Pyramid and about two times the total mass of American humanity.

STEP 22

31 600 000 000 kilograms ($10^{10.5}$ kg)
31.6 megatonnes ($10^{1.5}$ Mt)

The population of India has a total mass of about 30 megatonnes, or 1.5 times the mass of concrete in the Grand Coulee Dam.
If we consider that portion of Earth's atmosphere that rests on the United States, it contains about 31.6 megatonnes of the rare noble gas xenon.

STEP 23

100 000 000 000 kilograms (10^{11} kg)
100 megatonnes (10^{2} Mt)

The population of Asia has a total mass of about 100 megatonnes, five times the mass of concrete in the Grand Coulee Dam.
The rarest types of stable atoms on Earth are helium-3 and xenon-126, each of which is found only in the atmosphere and each of which makes up only a small proportion of the total mass of the gases (themselves rare) of which they are varieties—helium and xenon. Even so, the total amount of helium-3 in the Earth's atmosphere

amounts to about 100 megatonnes, while the amount of xenon-126 is about 200 megatonnes.

STEP 24

316 000 000 000 kilograms ($10^{11.5}$ kg)
316 megatonnes ($10^{2.5}$ Mt)

Having reached the Step-24 mass-range, we have outpaced the total mass of humanity. The entire population of the Earth, which is presently about 4 400 000 000, has a combined mass of about 220 megatonnes, if we count the average mass of a human being as 50 kilograms.

This is only about 11 times the mass of the Grand Coulee Dam, and this would make it seem that the encumbrance of humanity upon the Earth is not very great. After all, the Earth could easily support eleven dams the size of the Grand Coulee—why may it not as easily support the equivalent mass of human flesh and blood?

Consider, however, the quantity of resources consumed, and wastes produced, by the human mass. Humanity, you see, is not a passive mass resting on the Earth, as the Grand Coulee Dam is; humanity is engaged in active procedures that, especially in these industrialized times, place an enormous strain on the planet.

The smallest astronomical object that has received a name is the asteroid Hermes, which approached Earth very closely in 1937. Its diameter has been estimated at 600 metres, and assuming it to be roughly spherical and to be composed of stony material, its mass would be a little over 300 megatonnes.

STEP 25

1 000 000 000 000 kilograms (10^{12} kg)
1 gigatonne (10^{0} Gt)

At the Step-25 mass-range, we reach the largest single construction effort of the human species—and it is preindustrial.

In the third century B.C., the Chinese began to build a wall across the northern boundary of China in order to keep out the marauding nomads of Central Asia. (It couldn't keep them out, but it could keep out their horses, and without their horses, they could be dealt with.) The Great Wall, as it is called, was kept in repair, strengthened, and improved for 17 centuries. It ended up being about 6.3 megametres long (counting all its branches and spurs), and, on the average, about 9 metres high and 10 metres thick.

This would give it a total mass of perhaps 1.5 gigatonnes, which is 5 times that of the asteroid Hermes, nearly 7 times the mass of the entire human species, and about 240 times the mass of the Great Pyramid.

However, while Hermes and the Great Pyramid have constant masses, humanity as a species does not. The total mass of humanity is increasing steadily, and at the present rate of increase, it will be equal to that of the Great Wall in about 100 years. It is doubtful whether the Earth can feed and otherwise support so large a mass of human beings so that, unless a rational population policy is adopted, humanity may be facing utter disaster before very many decades have passed.

STEP 26

3 160 000 000 000 kilograms ($10^{12.5}$ kg)
3.16 gigatonnes ($10^{0.5}$ Gt)

The total amount of atmospheric xenon (in all its varieties) is equal to about 2 gigatonnes, while that of helium is about 3.6 gigatonnes.

If the atmospheric supply of helium were all there were on Earth, that would be indeed unfortunate, because helium has a number of important uses in science and technology, and it is irreplaceable, for there is nothing else with its properties. To be sure, 3.6 gigatonnes sounds like a great deal, being 2.4 times the mass of the Great Wall of China; but that mass is spread throughout the Earth's atmosphere, so that it would be all but impossible to concentrate and extract that helium without prohibitive expense.

Fortunately, helium is produced in the course of the radioactive breakdown of uranium and thorium so that there are quantities of it trapped underground that have been accumulating over vast periods of time. Considerable supplies can be obtained from some wells which produce natural gas, primarily.

STEP 27

10 000 000 000 000 kilograms (10^{13} kg)
10 gigatonnes (10^{1} Gt)

The asteroid Icarus has a diameter of about 1.4 kilometres. Assuming it to be roughly spherical, and stony in nature, it would have a mass of about 8 gigatonnes.

The outermost layer of the Earth's solid globe is the crust. It is a very thin layer, only 17 kilometres thick, on the average, and that is very little compared to the total thickness of the globe. It is from the crust that we obtain our mineral resources (indeed, from the upper portion of the crust).

Geologists have estimated the average percentage composition of the crust in terms of the elements that make it up, and from this we can calculate the total mass of particular elements within the crust. For instance, among the rarest of the solid elements is iridium, an element very much like the more familiar platinum in its properties. The total amount of iridium in the Earth's crust is about 24 gigatonnes.

This means that if we could get all the iridium out of the crust (a very difficult undertaking indeed and, in fact, one that is totally impractical at the present time) we could, in imagination, make of it an asteroid about three times as massive as Icarus. However, iridium is more massive, size for size, than the material that probably makes up Icarus, so that an iridium asteroid 24 gigatonnes in mass would have a diameter of only about 1 kilometre.

STEP 28

31 600 000 000 000 kilograms ($10^{13.5}$ *kg*)
31.6 gigatonnes ($10^{1.5}$ *Gt*)

The total quantity of helium in the Earth's crust may have a mass of 72 gigatonnes, or 20 times the quantity of helium in the Earth's atmosphere. This makes it clear why the crust is a better source of helium than the air is, especially since in the air the helium spreads out evenly, whereas in the crust it tends to be ready-concentrated in places where the uranium and thorium ores happen to be located.

Helium is constantly being lost to the atmosphere as natural gas is mined and burned, and even when it is extracted and used, it eventually leaks into the atmosphere, and then away into outer space.

Helium is the one resource Earth can really *lose.* It is not just a matter of using it, and then throwing it away into garbage heaps or sewers from which it might conceivably be recovered; or changing it chemically, in ways which might conceivably be reversed. Helium is slowly lost to outer space, from which recovery is an entirely different matter.

STEP 29

100 000 000 000 000 kilograms (10^{14} kg)
100 gigatonnes (10^{2} Gt)

The total amount of gold in the Earth's crust is about 120 gigatonnes. The total amount of platinum there is about the same. All this gold, at the current market prices, would have a value of a quarter of a quadrillion dollars, and would be enough to pay the American national debt two hundred fifty times over.

There are two catches. First, it is completely impractical to hope to get out more than a small fraction of the gold distributed through the Earth's crust. In all the history of gold-mining, only one millionth of the total supply has been extracted. Second, the chief reason why gold is so valuable is that it is so rare. If all the gold in the crust were to be available, the price of gold would drop to virtually nothing.

STEP 30

316 000 000 000 000 kilograms ($10^{14.5}$ kg)
316 gigatonnes ($10^{2.5}$ Gt)

Uranium, as it occurs in nature, consists of two varieties or "isotopes." Of these, the less common is uranium-235, which is particularly important because it is easily subject to neutron-induced fission and is therefore the fundamental fuel for both bombs and for peaceful nuclear power. The total quantity of uranium-235 in Earth's crust is about 345 gigatonnes.

Leda, the smallest of the outer satellites of Jupiter, may be in this mass-range, too.

STEP 31

1 000 000 000 000 000 kilograms (10^{15} kg)
1 teratonne (10^{0} Tt)

Now we are in the teratonne range. Deimos, the outer and smaller satellite of Mars, has a mass of about 1.5 teratonnes, or nearly 200 times the mass of Icarus.

This is still small on the Earthly scale. Carbon dioxide is a minor constituent of the Earth's atmosphere, making up only 3.2 parts per ten thousand. Even so, the total quantity of carbon dioxide in the Earth's atmosphere is about 2.5 teratonnes, 1.7 times the mass of Deimos. Atmospheric carbon dioxide is about equal to the amount of silver in the Earth's crust.

STEP 32

3 160 000 000 000 000 kilograms ($10^{15.5}$ kg)
3.16 teratonnes ($10^{0.5}$ Tt)

The total amount of antimony in the Earth's crust is about 5 teratonnes. This is also about the total mass of Eros, the largest and best known of the Earth-grazing asteroids.

STEP 33

10 000 000 000 000 000 kilograms (10^{16} kg)
10 teratonnes (10^{1} Tt)

The innermost of Mars's satellites, Phobos, is just about 10 teratonnes in mass, exactly at the Step-33 mass-range.

Phobos is a lump of rock about 22 kilometres in diameter, on the average. Compare it to the Martian atmosphere, which is only 1/200 as massive as the Earth's, quantity for quantity. What's more, the Martian atmosphere is spread out over an area only one third that over which Earth's atmosphere is spread out. Even so, the mass of Mars's atmosphere is about 22 teratonnes or just about twice that of Phobos and Deimos put together.

Now let us take a last look at life. The mass of all living things on Earth (the biosphere) is estimated, rather shakily, to be about 17 teratonnes. This is nearly 80 000 times the total mass of humanity.

This sounds as though human beings make up very little of all of life on Earth, but, to the contrary, no large species has ever, in all Earth's history, made up so large a fraction of the total mass of life. At the present rate at which the human population is increasing (if it maintains that rate), then in 650 years, the mass of humanity will be equal to the total quantity of life now present.

STEP 34

31 600 000 000 000 000 kilograms ($10^{16.5}$ kg)
31.6 teratonnes ($10^{1.5}$ Tt)

The total amount of tungsten (the metal out of which we make electric light filaments) in the Earth's crust is equal to about 24 teratonnes.

The ocean is almost entirely water, which is a compound of hydrogen and oxygen. Of this, hydrogen makes up a ninth of the mass, and almost all of it is the isotope hydrogen-1. A small fraction (about 1 out of 7 000) is made up of hydrogen-2, or "deuterium." The total quantity of deuterium in the ocean is 21.6 teratonnes, equal to something less than the total quantity of tungsten in the crust.

However, while only a small portion of the tungsten can be extracted from the crust, most of the deuterium can be extracted from the ocean (in time) without too much trouble. In view of the fact that enormous quantities of energy can be obtained from nuclear fusion of deuterium (once the proper method for producing and controlling the energy can be worked out), the oceans offer us an energy source that could last for billions of years.

In contrast, the quantity of uranium in the Earth's crust (of both varieties) is about 48 teratonnes, more than twice the quantity of deuterium in the oceans, and all of it can be used for power in breeder reactors. However, mass for mass, considerably more energy can be obtained from deuterium than from uranium, and the deuterium can be obtained *much* more easily. Deuterium remains the better energy source, at least potentially.

The mass of argon in the Earth's atmosphere comes to a total of about 66 teratonnes. It is by far the most common of the noble gases. It is unusually common because an isotope of the element potassium (potassium-40) undergoes a slow radioactive breakdown, producing argon-40 among its breakdown products. This argon slowly leaks out of the soil into the atmosphere and has been accumulating there all through the history of the Earth.

The total amount of potassium-40 still present in the crust today is about 74 teratonnes, so that eventually, when the Earth is several billions of years older than it is now, the quantity of argon in the atmosphere will have doubled.

STEP 35

100 000 000 000 000 000 kilograms (10^{17} kg)
100 teratonnes (10^{2} Tt)

The amount of thorium in the Earth's crust is about 240 teratonnes. This is about five times the mass of uranium there, and since thorium, too, can serve as a basic fuel for nuclear fission, it increases the potential of fission as an energy source.

STEP 36

316 000 000 000 000 000 kilograms ($10^{17.5}$ kg)
316 teratonnes ($10^{2.5}$ Tt)

A Step-36 mass represents just about the total quantity of potentially fissionable material in the Earth's crust.

At this mass-range, we have also reached nearly the amount of liquid fresh water on Earth; the water in lakes, ponds, rivers, streams, and wells: the only kind of water that can be used directly for drinking, cooking, washing, as well as for agriculture and for industry. The total comes to about 500 teratonnes.

This would seem to be plenty since it represents 100 kilotonnes for each man, woman, and child on Earth. However, this water is not evenly distributed, and while some portions of the inhabited Earth have sufficient liquid fresh water and even more than sufficient at times, other portions have a chronic shortage. Furthermore, what liquid fresh water exists is being increasingly polluted, thanks to the carelessness and indifference of human beings, so that the world, before long, may face a general shortage of this absolutely vital resource.

STEP 37

1 000 000 000 000 000 000 kilograms (10^{18} kg)
1 petatonne (10^{0} Pt)

At the petatonne range, we now reach one of the major components of the atmosphere. The total mass of atmospheric oxygen is 1.19 petatonnes. This is also about the mass of Himalia, the largest of Jupiter's outer satellites.

STEP 38

3 160 000 000 000 000 000 kilograms ($10^{18.5}$ kg)
3.16 petatonnes ($10^{0.5}$ Pt)

The total mass of nitrogen in our atmosphere is 3.88 petatonnes. It is the largest component of the atmosphere, making up three fourths of its mass. The total mass of Earth's atmosphere is 5.136 petatonnes. This is roughly equal to the total mass of chromium in the Earth's crust.

STEP 39

10 000 000 000 000 000 000 kilograms (10^{19} kg)
10 petatonnes (10^{1} Pt)

The total mass of ice on the Earth is about 22.8 petatonnes. Ice is frozen fresh water and, if melted, could add quite substantially to the fresh water supplies of the world, for the total mass of ice on Earth is about 45 times that of liquid fresh water. However, ice is not, on the whole, very accessible. About nine tenths of it (20 petatonnes) makes up the vast ice sheet that covers Antarctica, and most of the rest makes up the smaller ice sheet that covers Greenland.

The total amount of sulfur in the Earth's crust is 12.5 petatonnes, only about five eighths the mass of the Antarctica ice sheet.

STEP 40

31 600 000 000 000 000 000 kilograms ($10^{19.5}$ kg)
31.6 petatonnes ($10^{1.5}$ Pt)

In this mass-range, we are beginning to reach the larger asteroids and the medium-sized satellites. Thus, Miranda, the innermost and smallest of the satellites of Uranus, has a mass of about 34 petatonnes, while Mimas, the innermost sizable satellite of Saturn, has a mass of about 37 petatonnes. The large asteroid Davida, which is 260 kilometres in diameter, has a mass of about 30 petatonnes.

All the solid matter dissolved in Earth's ocean (by far the major portion of which is ordinary salt) has a mass of 47 petatonnes. In other words, if the ocean solids were gathered into a spherical mass, it would be 340 kilometres in diameter, and would make an asteroid nearly as large as the largest that actually exists.

STEP 41

100 000 000 000 000 000 000 kilograms (10^{20} kg)
100 petatonnes (10^{2} Pt)

I remember watching a television science fiction program in which colonists on Earth's Moon are running low on the "rare metal" titanium. They must undergo enormous difficulties to find the small quantity of it they need.

I could only shake my head sadly. Titanium, in actual fact, is the ninth most common element in the Earth's crust, and the Moon's crust is even richer in it. The Earth's crust contains just about 100 petatonnes of titanium.

Saturn's satellite Enceladus, which lies just beyond Mimas, has a mass of about 74 petatonnes, so there is enough titanium in the Earth's crust to build a satellite 1.3 times as massive as Enceladus.

STEP 42

316 000 000 000 000 000 000 kilograms ($10^{20.5}$ *kg*)
316 petatonnes ($10^{2.5}$ *Pt*)

The total amount of magnesium in the Earth's crust, where it is the eighth most common element, is 500 petatonnes. Compare this with Tethys, one of Saturn's satellites, which has a mass of 626 petatonnes.

Venus's atmosphere is much denser than Earth's and, in total, has a mass of 417 petatonnes, better than 81 times that of Earth's.

STEP 43

1 000 000 000 000 000 000 000 kilograms (10^{21} *kg*)
1 exatonne (10^{0} *Et*)

At the Step-43 mass-range, we have reached nearly the mass of the water in Earth's ocean, which is about 1.3 exatonnes.

The ocean contains not quite all of Earth's water supply. There remains, besides, the quantity in the ice caps, in the liquid fresh water supply, even in the water vapor in the atmosphere. Add it all together and the total water supply of Earth, or "hydrosphere," has a mass of 1.37 exatonnes. This is a mass that is larger than that of Titania, Uranus's largest satellite, which has a mass of about 1.2 exatonnes. What's more, Ceres, the largest asteroid, has a mass of only about 1 exatonne, less than three fourths the mass of Earth's total water supply.

If all the water on Earth were collected into a huge sphere, it would be 1.4 megametres in diameter. If such a sphere revolved about the Earth at a distance of 150 megametres (five eighths the distance of the Moon) it would look as large as the Moon in our sky.

Iron is the fourth most common element in the Earth's crust. All the iron in the Earth's crust has a mass of 1.2 exatonnes.

STEP 44

3 160 000 000 000 000 000 000 kilograms ($10^{21.5}$ *kg*)
3.16 exatonnes ($10^{0.5}$ *Et*)

Aluminum is the third most common element in Earth's crust, and the most common metal. The total mass of the crust's aluminum is equal to about 2 exatonnes.

The mass of Rhea, Saturn's second largest satellite, is estimated to be about 2.3 exatonnes.

STEP 45

10 000 000 000 000 000 000 000 kilograms (10^{22} kg)
10 exatonnes (10^{1} Et)

Silicon is the second most common element in Earth's crust. The total mass of the crust's silicon is equal to about 6.6 teratonnes. The only element more common is oxygen, which is not present in the crust in gaseous form, of course, but in combination with silicon and the various metals to form the stony "silicates" that make up the crust. The total mass of the crust's oxygen is just about 11.2 exatonnes.

STEP 46

31 600 000 000 000 000 000 000 kilograms ($10^{22.5}$ kg)
31.6 exatonnes ($10^{1.5}$ Et)

The total mass of the Earth's crust is 24 exatonnes. Add to that the mass of the ocean and atmosphere and it comes to nearly 25.5 exatonnes, not far short of the Step-46 mass.

We are now almost in the range of the giant satellites. Europa, the smallest of Jupiter's four large satellites, has a mass of 48.73 exatonnes, just about twice that of Earth's crust. Triton, Neptune's large satellite, has an estimated mass of 57 exatonnes.

STEP 47

100 000 000 000 000 000 000 000 kilograms (10^{23} kg)
100 exatonnes (10^{2} Et)

We are now in the mid-range of the giant satellites. In fact, the Step-47 mass somewhat exceeds that of our Moon, which has a mass of 73.5 exatonnes. Io, the innermost of the Galilean satellites, has a mass of 89.16 exatonnes. The three most massive satellites of the Solar system—Callisto, Titan, and Ganymede—have masses, respectively, of 107, 136, and 149 exatonnes. The most massive of all known satellites, Ganymede, has, therefore, a mass that is just twice that of our Moon.

Since we are now well past the mass of the Earth's crust, we must begin to consider the entire planet. Working out the mass of various elements over the Earth taken as a whole is an uncertain procedure and the results are very tentative. Based, however, on the best that can be done, we can say that the total mass of sulfur in the whole

Earth is about 160 exatonnes. In other words, all the sulfur in Earth, gathered together, would make a sphere that was more massive than Ganymede.

STEP 48

316 000 000 000 000 000 000 000 kilograms ($10^{23.5}$ kg)
316 exatonnes ($10^{2.5}$ Et)

Although Mercury is the smallest of the planets, and is actually smaller in diameter than Ganymede, for instance, it is made of materials that are denser than those making up Ganymede. Mercury has a mass that is 331 exatonnes, or just above the Step-48 level. Mercury is thus nearly 2 1/4 times as massive as Ganymede.

STEP 49

1 000 000 000 000 000 000 000 000 kilograms (10^{24} kg)
1 000 exatonnes (10^{3} Et)
(13.6 MN)

We have run out of prefixes accepted by the SI version of the metric system. In order to avoid having to use figures too large, we shall adopt the mass of the Moon (MN) as an additional unit, though I will keep listing exatonnes, too, as we move up the ladder of mass. Thus, 1 000 exatonnes is a mass that is equal to almost exactly 13.6 times that of the Moon (13.6 MN).

Mars, the second smallest of the planets, has a mass of 642 exatonnes (well below the Step-49 mass), and this is equal to 8.73 times the mass of the Moon (8.73 MN) or nearly twice the mass of Mercury.

Earth is divided into two main regions. At the very center, there is thought to be a metallic core, largely molten, composed of iron and nickel in a ten-to-one ratio. This core has a mass of about 1 875 exatonnes (25.5 MN). Earth's core alone is therefore three times as massive as all of Mars. Around the core is a rocky "mantle," which is considerably more massive than the core. Together, the core and mantle make up 99.6 percent of the Earth, the remaining 0.4 percent being the crust, the ocean, and the atmosphere.

Not only does iron make up the major portion of Earth's core, it is also well represented in the mantle and crust. All told, the quantity of iron in the Earth comes to 2 115 exatonnes (28.8 MN) or 3.3 times the mass of Mars.

STEP 50

3 160 000 000 000 000 000 000 000 kilograms ($10^{24.5}$ kg)
3 160 exatonnes ($10^{3.5}$ Et)
(43 MN)

Earth's mantle, which makes up two thirds of the Earth's mass, has a mass of about 4 080 exatonnes (55.5 MN).

STEP 51

10 000 000 000 000 000 000 000 000 kilograms (10^{25} kg)
10 000 exatonnes (10^{4} Et)
(136 MN) or (1.67 EA)

At this level, we have finally overstepped the entire Earth, which has a mass of 5 976 exatonnes (81.3 MN). The Step-51 level of 10 000 exatonnes represents a mass 1.67 times that of Earth, and we may call it 1.67 EA. We will now therefore switch from the Moon to the Earth for comparisons.

Venus, for instance, has a mass of 4 870 exatonnes (0.815 EA) and is thus only five sixths as massive as Earth. The entire planet of Venus is, in fact, only 1.2 times as massive as Earth's mantle.

If we take the mass of Earth and Venus together, that mass comes to 10 846 exatonnes (1.815 EA) and just tops the Step-51 mass-range.

As a matter of fact, if we take all the worlds of the inner Solar system—Mercury, Venus, Earth, Moon, Mars, plus the tiny worlds such as the Martian satellites, occasional asteroids, meteoroids, and comets—the total mass comes to about 11 900 exatonnes (1.99 EA). In other words, our planet, the Earth, has in its substance almost exactly half of all the mass of the inner Solar system (provided we don't count the Sun, of course).

Suppose, next, that we take *all* the Solar system, except for the Sun and the four giant planets. The total mass of the smaller planets, plus all the satellites, asteroids, comets, and assorted debris throughout the Solar system, comes to perhaps 12 500 exatonnes (2.09 EA). The additional minor items of the outer Solar system add a mass equal to one tenth that of the Earth, or about eight times that of the Moon.

If we leave out the Sun and the four giant planets, then, 47.8 percent of all the matter in the Solar system is concentrated in the Earth; and 86.8 percent, nearly seven eighths, of all the matter in the Solar system is concentrated in Earth and Venus.

STEP 52

31 600 000 000 000 000 000 000 000 kilograms ($10^{25.5}$ kg)
31 600 exatonnes ($10^{4.5}$ Et)
(5.27 EA)

For the first time, we are at a mass-level that cannot be easily exemplified. There are no objects or simple combinations of objects in the Solar system that come near a mass equal to 5 1/4 times that of the Earth. Nor are the structures of objects considerably more massive than the Earth so well known that we can take some easily expressed portion of them for our purposes. It may well be that there are objects outside the Solar system that are in this mass-range (planets circling other stars, as an example) but none of these are known to us. We must move on.

STEP 53

100 000 000 000 000 000 000 000 000 kilograms (10^{26} kg)
100 000 exatonnes (10^{5} Et)
(16.7 EA)

Now we move into the realm of the four giant planets. The least of these is Uranus, which has a mass of 86 900 exatonnes (14.54 EA) and is just short of the Step-53 mass-level. Neptune, with a mass of about 103 000 exatonnes (17.23 EA), is just above it.

STEP 54

316 000 000 000 000 000 000 000 000 kilograms ($10^{26.5}$ kg)
316 000 exatonnes ($10^{5.5}$ Et)
(52.7 EA)

The mass of Uranus and Neptune, together with Earth and Venus and all still smaller bodies in the Solar system, comes to 202 000 exatonnes (33.86 EA). This is the total mass of the Solar system, if the Sun, Jupiter and Saturn are subtracted, and it is still only two thirds of the Step-54 mass-level. At this stage, 93.8 percent of all the mass is in Neptune and Uranus, while the remaining 6.2 percent is in Earth, Venus, and the smaller bodies.

Neptune alone possesses as much mass as do all the bodies of the Solar system that are smaller than itself, including Uranus.

STEP 55

1 000 000 000 000 000 000 000 000 000 kilograms (10^{27} kg)
1 000 000 exatonnes (10^{6} Et)
(167 EA)

Saturn, the second largest planet, has a mass of about 570 000 exatonnes (95.15 EA), which is just about three times the mass of Uranus and Neptune taken together, but is still well short of the Step-55 mass-level.

If you add to Saturn all the objects of the Solar system that are smaller than itself, the total mass is 771 000 exatonnes (129 EA), and that is still only a little over three fourths of the Step-55 level. If the Sun and Jupiter are excluded, Saturn possesses three fourths of the total mass of the rest of the Solar system.

STEP 56

3 160 000 000 000 000 000 000 000 000 kilograms ($10^{27.5}$ kg)
3 160 000 exatonnes ($10^{6.5}$ Et)
(527 EA or 1.66 JU)

The largest planet, Jupiter, has a mass of about 1 900 000 exatonnes (317.9 EA). The Step-56 mass-level is equal to 1.66 times the mass of Jupiter (1.66 JU), therefore.

Everything in the Solar system except for the Sun itself has a mass of 2 670 000 exatonnes (1.4 JU). Jupiter thus contains 71 percent of all the mass of the Solar system, excluding the Sun. The four giant planets together make up 99.5 percent of the mass of the Solar system excluding the Sun. For this reason, some extraterrestrial visitor, viewing our Solar system from afar, might well record in his report that the Sun was circled by four planets plus debris.

STEP 57

10 000 000 000 000 000 000 000 000 000 kilograms (10^{28} kg)
10 000 000 exatonnes (10^{7} Et)
(5.25 JU)

Five steps previously we could find nothing that would represent a mass of about five times that of Earth; now we can find nothing known that would represent a mass of about five times that of Jupiter.

On this occasion, however, there is a hint of something.

Astronomers have been searching for some indication that stars other than our Sun might have planetary systems of their own. It seems ridiculous to suppose that they haven't and that our Sun is unique among so many, but it would be nice to obtain direct evidence for extra-Solar planetary systems.

So far, the best that astronomers can do is to study stars whose pathway through space might wobble slightly because of the gravitational pull of some planet orbiting about it. This would be just barely detectable if the star was small, the planet large, and both were relatively close to the Earth.

Some nearby small stars have presented wobbles just barely within the limit of visibility, and there is a chance that this might mean that such extra-Solar planets have been detected. In some cases, it has been estimated that the planet detected is equal in mass to about 15 000 000 exatonnes (8 JU). If so, such extra-Solar super-Jupiters are objects just beyond the Step-57 mass-level.

STEP 58

31 600 000 000 000 000 000 000 000 000 kilograms ($10^{28.5}$ kg)
31 600 000 exatonnes ($10^{7.5}$ Et)
(16.6 JU)

The more massive an object is, the hotter it is at the core and the greater the pressures there. Jupiter is not quite massive enough to possess the central pressures and temperatures necessary for igniting nuclear fusion and setting the planet to glowing like a star. Even the super-Jupiters that may be circling other stars might not be large enough.

At some mass-level not very much higher than that of Jupiter, however, an object will indeed graduate to star-level. There is a very good chance that the Step-58 mass-level is at or near the dividing line between planets and stars. However, we have no clear examples of objects at this level.

This is not surprising. An object with a mass at this level would, if it ignites, be dimly red-hot. It would glow feebly indeed and would barely be visible even if it was very close to us (as stellar distances go).

STEP 59

100 000 000 000 000 000 000 000 000 000 kilograms (10^{29} kg)
100 000 000 exatonnes (10^{8} Et)
(52.5 JU)

At this level, there may be actual stars visible. At least, there is a dim red star listed in the catalogues as Luyten 726-8 B that may be no more than 75 000 000 exatonnes (40 JU) in mass.

STEP 60

316 000 000 000 000 000 000 000 000 000 kilograms ($10^{29.5}$ kg)
316 000 000 exatonnes ($10^{8.5}$ Et)
(166 JU)

We have now moved to a level of unmistakable stars. The nearest star of all, Proxima Centauri, is a dim red star with a mass of perhaps 430 000 000 exatonnes (230 JU).

Proxima Centauri is presumably part of the Alpha Centauri system, along with two bright stars, Alpha Centauri A and Alpha Centauri B. The nearest star that is not part of that system is Barnard's Star, and it, too, is perhaps 430 000 000 exatonnes (230 JU) in mass.

STEP 61

1 000 000 000 000 000 000 000 000 000 000 kilograms (10^{30} kg)
1 000 000 000 exatonnes (10^{9} Et)
(525 JU)

We may now deal with fairly bright stars. The star 61 Cygni B is part of a two-star system of which the other is the somewhat brighter 61 Cygni A. The two together were the first stars to have their distances determined. 61 Cygni B has a mass of 1 300 000 000 exatonnes (680 JU), while 61 Cygni A has a mass of 1 400 000 000 exatonnes (730 JU).

STEP 62

3 160 000 000 000 000 000 000 000 000 000 kilograms ($10^{30.5}$ kg)
3 160 000 000 exatonnes ($10^{9.5}$ Et)
(1 660 JU or 1.58 SU)

The two stars of the 61 Cygni system have, together, a mass of 2 700 000 000 exatonnes (1 410 JU), and, as a system, are nearly at the Step-62 mass-level.

It is here that we finally pass the mass of the Sun. The Sun has a mass of 1 989 000 000 exatonnes (1 047 JU). The Step-62 mass-level of 3 160 000 000 exatonnes is equal to 1.58 times that of the Sun (or 1.58 SU) so we will now shift from Jupiter to the Sun as a mass reference.

The Sun, with its mass of 1 047 JU, has a mass just about 750 times as great as that of all the planetary matter (including Jupiter) that circles round it. It has nearly 99.9 percent of the mass of the entire Solar system. No wonder it is hard to detect planets circling other stars, at their huge distances, if anything near this disproportion exists in the cases of stars generally (and it probably does). The situation is made even worse by the fact that planets are so close to stars, and that planets shine only by reflected light and are, therefore, so much dimmer than the stars they circle.

Alpha Centauri A has a mass almost exactly equal to that of the Sun. If the three stars of the Alpha Centauri system are taken together, the total mass is just under 4 000 000 000 exatonnes (2 SU).

The white dwarf star, Sirius B, that circles Sirius, has a mass almost exactly that of the Sun, too, even though its diameter and volume are substantially less than the Earth's.

STEP 63

10 000 000 000 000 000 000 000 000 000 000 kilograms (10^{31} kg)
10 000 000 000 exatonnes (10^{10} Et)
(5 SU)

We have lifted ourselves well beyond the Sun now. The bright star, Sirius, has a mass of about 5 000 000 000 exatonnes (2.5 SU). Add to that its companion star, Sirius B, and the total mass of the system is 7 000 000 000 exatonnes (3.5 SU), which is still well below the Step-63 mass-level. For a single star with a mass in the Step-63 range, there is the bright star Achernar.

STEP 64

31 600 000 000 000 000 000 000 000 000 000 kilograms ($10^{31.5}$ kg)
31 600 000 000 exatonnes ($10^{10.5}$ Et)
(15.8 SU)

Although there are giant red stars with enormous volumes, which we mentioned in the ascending ladder of volume, those volumes consist almost entirely of very thin gas. The total mass doesn't increase by more than small fractions of the volume increase.

Stars at the Step-64 mass-level are about as massive as they can be, for the more massive a star, the hotter it must be if it is to maintain its expanded volume against the increasing gravitational pull inward. If stars are more massive still, the necessary balance between internal heat and gravitational pull cannot be maintained, and the star either explodes, or collapses, or both.

At the Step-64 mass-level, we have the stars of the "O spectral class," the hottest, brightest, and shortest-lived of the stars (at least, shortest-lived in the sense of a steadily shining star of the kind we are familiar with in the case of the Sun). Only one star in ten million belongs to this class, and there may not be more than a few tens of thousands of them in the entire Galaxy.

STEP 65

100 000 000 000 000 000 000 000 000 000 000 kilograms (10^{32} *kg)*
100 000 000 000 exatonnes (10^{11} *Et)*
(50 SU)

The most massive stars that are thought to be a pair (Plaskett's stars, named for the discoverer) circle each other and are each about 140 000 000 000 exatonnes (70 SU) in mass. There may be stars more massive still, but their existence has not yet been definitely established.

STEP 66

316 000 000 000 000 000 000 000 000 000 000 kilograms ($10^{32.5}$ *kg)*
316 000 000 000 exatonnes ($10^{11.5}$ *Et)*
(158 SU)

At the Step-66 mass-level, we leave individual stars behind and can, henceforth, speak only of star systems. The Plaskett binary, for instance, has a total mass of 280 000 000 000 exatonnes (140 SU). We might also speak of open clusters, associations of numerous stars bound gravitationally, but not very closely. An open cluster, such as the Pleiades, may have a total mass of as much as 400 000 000 000 exatonnes (200 SU).

STEP 67

1 000 000 000 000 000 000 000 000 000 000 000 kilograms (10^{33} *kg)*
1 000 000 000 000 exatonnes (10^{12} *Et)*
(500 SU)

The Hyades, an open cluster that is larger than the Pleiades (both are in the constellation of Taurus), may have a total mass of up to 800 000 000 000 exatonnes (400 SU).

STEP 68

3 160 000 000 000 000 000 000 000 000 000 000 kilograms ($10^{33.5}$ kg)
3 160 000 000 000 exatonnes ($10^{12.5}$ Et)
(1 580 SU)

The largest open clusters may have total masses of up to 4 000 000 000 000 exatonnes (about 2 000 SU).

STEP 69

10 000 000 000 000 000 000 000 000 000 000 000 kilograms (10^{34} kg)
10 000 000 000 000 exatonnes (10^{13} Et)
(5 000 SU)

In addition to the open clusters, there are "globular clusters." These are far richer in stars than the open clusters are, and their stars are tightly packed into a globular structure (hence the name). There are about 125 globular clusters known in our Galaxy, and there may be at least as many more that we don't see because they are hidden by dust clouds. The smallest globular clusters may have total masses of not much more than 10 000 000 000 000 exatonnes (5 000 SU).

STEP 70

31 600 000 000 000 000 000 000 000 000 000 000 kilograms ($10^{34.5}$ kg)
31 600 000 000 000 exatonnes ($10^{13.5}$ Et)
(15 800 SU)

M71, a small globular cluster in the constellation Sagitta, may have a total mass of about 30 000 000 000 000 exatonnes (15 000 SU).

STEP 71

100 000 000 000 000 000 000 000 000 000 000 000 kilograms (10^{35} kg)
100 000 000 000 000 exatonnes (10^{14} Et)
(50 000 SU)

M3, a globular cluster in the constellation Canes Venatici, may have a total mass of about 120 000 000 000 000 exatonnes (60 000 SU).

STEP 72

316 000 000 000 000 000 000 000 000 000 000 000 kilograms ($10^{35.5}$ kg)
316 000 000 000 000 exatonnes ($10^{14.5}$ Et)
(158 000 SU)

M92, a globular cluster in the constellation Hercules, may have a total mass of about 280 000 000 000 000 exatonnes (140 000 SU).

STEP 73

1 000 000 000 000 000 000 000 000 000 000 000 000 kilograms (10^{36} kg)
1 000 000 000 000 000 exatonnes (10^{15} Et)
(500 000 SU)

The very largest globular clusters have total masses approaching 2 000 000 000 000 000 exatonnes (1 000 000 SU).

The globular clusters overlap galaxies. In general, galaxies are more massive than globular clusters, but the very smallest galaxies are smaller than the largest globular clusters. (You can tell them apart because globular clusters surround large galaxies while small galaxies are more nearly independent and are looser and more irregular in structure.) The smallest galaxies may have total masses of as little as 1 000 000 000 000 000 exatonnes (500 000 SU).

STEP 74

3 160 000 000 000 000 000 000 000 000 000 000 000 kilograms ($10^{36.5}$ kg)
3 160 000 000 000 000 exatonnes ($10^{15.5}$ Et)
(1 580 000 SU)

The Leo II system, one of the dwarf galaxies that is part of our Local Group, has a mass of about 2 000 000 000 000 000 exatonnes (1 000 000 SU).

STEP 75

10 000 000 000 000 000 000 000 000 000 000 000 000 kilograms (10^{37} kg)
10 000 000 000 000 000 exatonnes (10^{16} Et)
(5 000 000 SU)

The Leo I system, another dwarf galaxy in our Local Group, has a mass of about 10 000 000 000 000 000 exatonnes (5 000 000 SU).

STEP 76

31 600 000 000 000 000 000 000 000 000 000 000 000 kilograms ($10^{37.5}$ kg)
31 600 000 000 000 000 exatonnes ($10^{16.5}$ Et)
(15 800 000 SU)

The Fornax system, also in our Local Group, has a total mass of about 24 000 000 000 000 000 exatonnes (12 000 000 SU).

STEP 77

100 000 000 000 000 000 000 000 000 000 000 000 000 kilograms (10^{38} kg)
100 000 000 000 000 000 exatonnes (10^{17} Et)
(50 000 000 SU)

There are no nearby galaxies at this level of mass, but undoubtedly there are many in the Universe generally.

STEP 78

316 000 000 000 000 000 000 000 000 000 000 000 000 kilograms ($10^{38.5}$ kg)
316 000 000 000 000 000 exatonnes ($10^{17.5}$ Et)
(158 000 000 SU)

Galaxy IC1613 of the Local Group has a total mass of about 500 000 000 000 000 000 exatonnes (250 000 000 SU).

STEP 79

1 000 000 000 000 000 000 000 000 000 000 000 000 000 kilograms (10^{39} kg)
1 000 000 000 000 000 000 exatonnes (10^{18} Et)
(500 000 000 SU)

We have now reached the level of middle-sized galaxies. Galaxy NGC 147 of the Local Group has a total mass of about 2 000 000 000 000 000 000 exatonnes (1 000 000 000 SU).

STEP 80

3 160 000 000 000 000 000 000 000 000 000 000 000 000 kilograms ($10^{39.5}$ kg)
3 160 000 000 000 000 000 exatonnes ($10^{18.5}$ Et)
(1 580 000 000 SU)

The Small Magellanic Cloud has a total mass of perhaps 4 000 000 000 000 000 000 exatonnes (2 000 000 000 SU).

STEP 81

10 000 000 000 000 000 000 000 000 000 000 000 000 kilograms (10^{40} kg)
10 000 000 000 000 000 000 exatonnes (10^{19} Et)
(5 000 000 000 SU)

The Large Magellanic Cloud, our nearest neighbor among the galaxies, has a total mass of nearly 20 000 000 000 000 000 000 exatonnes (10 000 000 000 SU).

STEP 82

31 600 000 000 000 000 000 000 000 000 000 000 000 kilograms ($10^{40.5}$ kg)
31 600 000 000 000 000 000 exatonnes ($10^{19.5}$ Et)
(15 800 000 000 SU)

Galaxy M33 has a total mass of about 26 000 000 000 000 000 000 exatonnes (13 000 000 000 SU).

STEP 83

100 000 000 000 000 000 000 000 000 000 000 000 000 kilograms (10^{41} kg)
100 000 000 000 000 000 000 exatonnes (10^{20} Et)
(50 000 000 000 SU)

We are now approaching the range of the giant galaxies.

STEP 84

316 000 000 000 000 000 000 000 000 000 000 000 000 kilograms ($10^{41.5}$ kg)
316 000 000 000 000 000 000 exatonnes ($10^{20.5}$ Et)
(158 000 000 000 SU or 1 MW)

The Step-84 mass-level represents almost exactly the estimated mass of our own Milky Way Galaxy. It is about 150 000 000 000 times as massive as our own Sun, so we can set 316 000 000 000 000 000 000 exatonnes as equal to 1 Milky Way Galaxy, or 1 MW.

STEP 85

1 000 000 000 000 000 000 000 000 000 000 000 000 000 000 kilograms (10^{42} kg)
1 000 000 000 000 000 000 000 exatonnes (10^{21} Et)
(3.16 MW)

Although our Milky Way Galaxy is a giant galaxy, it is by no means the largest giant. In our own Local Group, the Maffei galaxy (named for the astronomer who discovered it) has a mass of about 410 000 000 000 000 000 000 exatonnes (1.3 MW). The largest galaxy in our Local Group is the Andromeda Galaxy, which has a mass of about 730 000 000 000 000 000 000 exatonnes (2 MW). The entire Local Group has a mass of about 1 600 000 000 000 000 000 000 exatonnes (5 MW).

STEP 86

3 160 000 000 000 000 000 000 000 000 000 000 000 000 000 kilograms ($10^{42.5}$ kg)
3 160 000 000 000 000 000 000 exatonnes ($10^{21.5}$ Et)
(10 MW)

The Milky Way Galaxy and the Andromeda Galaxy are both spiral galaxies. They look like pinwheels when seen broadside, with a bright nucleus and with spiral arms. The most massive spirals are not much larger than the Andromeda Galaxy.

There is, however, another group, the "elliptical galaxies," which are relatively featureless and which look like monster globular clusters. They include the most massive galaxies, and there are a number of elliptical galaxies in the Step-86 mass-range.

STEP 87

10 000 000 000 000 000 000 000 000 000 000 000 000 000 000 kilograms (10^{43} kg)
10 000 000 000 000 000 000 000 exatonnes (10^{22} kg)
(31.6 MW)

The elliptical galaxy M87 is 8 000 000 000 000 000 000 000 exatonnes (25 MW) in mass, more or less, and we are now in the mass-range of sizable clusters of galaxies, as well. A cluster in the constellation Bootes contains as many as 150 galaxies (instead of the two dozen or so of our own Local Group), and it may have a total mass of about 10 000 000 000 000 000 000 000 exatonnes (30 MW).

STEP 88

31 600 000 000 000 000 000 000 000 000 000 000 000 000 000 kilograms ($10^{43.5}$ kg)
31 600 000 000 000 000 000 000 exatonnes ($10^{22.5}$ Et)
(100 MW)

The largest known elliptical galaxy may have a mass of as much as 20 000 000 000 000 000 000 000 exatonnes (60 MW), twice the size of the large cluster in Bootes, but that is still well short of the Step-88 mass-level. To reach that level, we must rely on clusters of galaxies. There is a cluster in the constellation Perseus, made up of 500 galaxies, that may have a total mass at the Step-88 level, for instance.

STEP 89

100 000 000 000 000 000 000 000 000 000 000 000 000 000 000 kilograms (10^{44} kg)
100 000 000 000 000 000 000 000 exatonnes (10^{23} Et)
(316 MW)

Here we must deal with giant clusters of over a thousand galaxies. One in the constellation of Virgo has perhaps 2 500 galaxies with a total mass of about 150 000 000 000 000 000 000 000 exatonnes (500 MW).

Beyond Step 89, there are no clear-cut examples of items to cite, merely more and more clusters of galaxies and clusters of clusters and clusters of clusters of clusters. Let us skip ahead 18 steps then and come at once to—

. . .

STEP 108

316 000 000 000 000 000 000 000 000 000 000 000 000 000 000 000 000 000 kilograms ($10^{53.5}$ kg)

316 000 000 000 000 000 000 000 000 000 000 exatonnes ($10^{32.5}$ Et)
(1 000 000 000 000 MW)

It is possible that there are 100 billion galaxies in the Universe altogether. In addition, there is the possibility of mass of other types —gaseous halos about the galaxies, undetected black holes,, vast numbers of neutrinos that are thought to be massless but may prove to have mass after all. It may be, then, that the Universe has a mass as high as 500 000 000 000 000 000 000 000 000 000 exatonnes (1 600 000 000 000 MW). The Step-108 mass-level should therefore be in the neighborhood of the maximum likely mass of the Universe, according to present-day notions.

(However, if neutrinos lack mass, and if other sources of mass, such as the black holes, also fail us, then the Universe may have a total mass closer to Step 107 or even Step 106.)

To summarize, then, we have gone from 1 kilogram (the mass of the smallest breed of dog) to the mass of the entire Universe in 107 steps, covering 53.5 orders of magnitude. It is time, now, to return to 1 kilogram and to follow the ladder of mass downward.

The Ladder of Mass

DOWNWARD

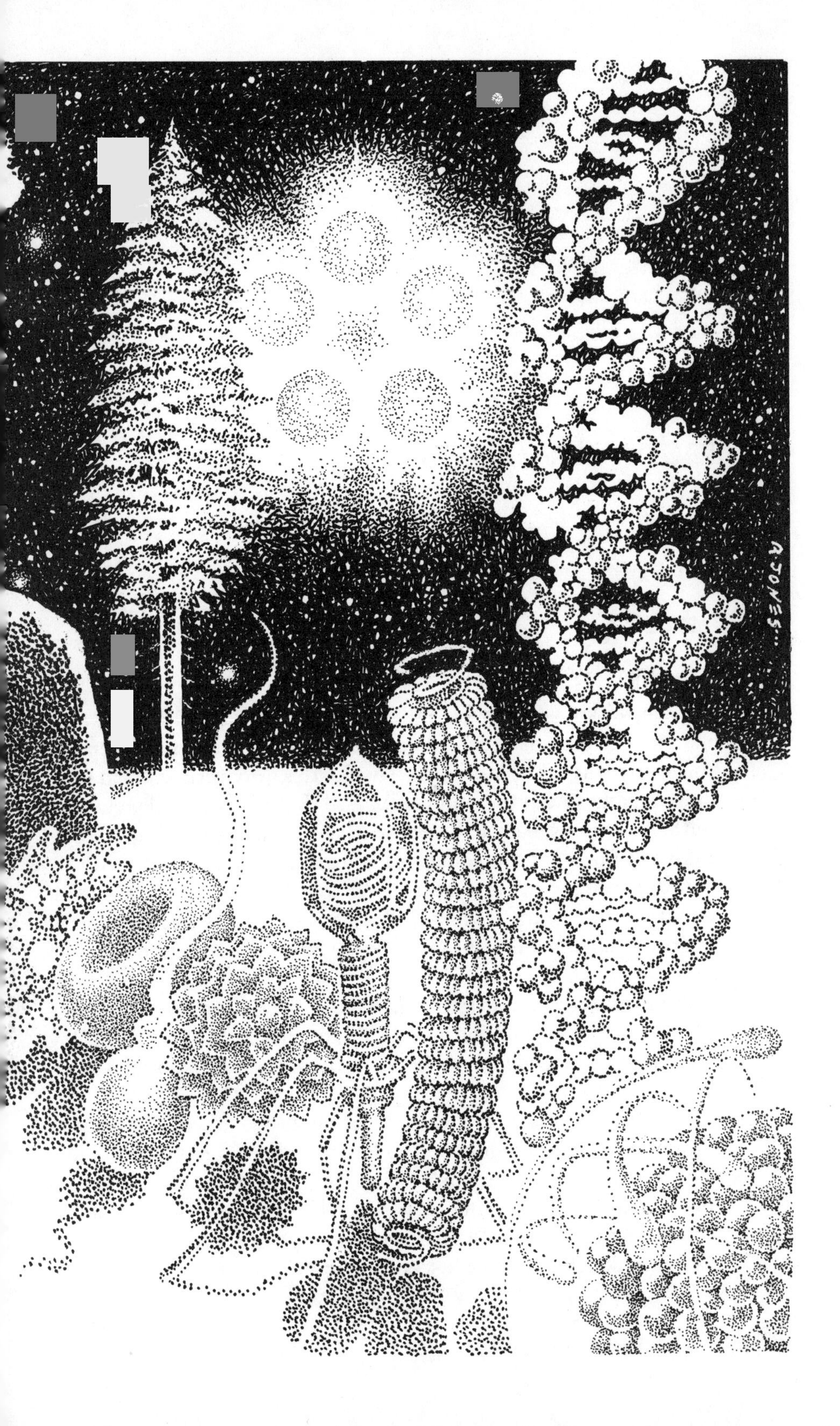
R. JONES

DOWNWARD

STEP 1

1 kilogram (10^0 kg)

How do we know when something has a mass of 1 kilogram? One method is to place it in one pan of a balance, and then place a known kilogram mass in the other pan. The Earth's gravitational field pulls on both pans, and if the pull is equal, so that the pans balance, with neither dropping below the other, the two masses are equal. The unknown mass is then equal to 1 kilogram also.

But how does one know that the "known" kilogram mass is itself actually 1 kilogram in mass? We assume it has been balanced against some standard—but how do we know the standard is 1 kilogram in mass?

The ultimate standard, to which, in theory, all masses are compared, is a cylinder of platinum-iridium (chosen because this alloy has a minimum tendency to chemical change) that is kept in a Paris suburb.

How large is such a standard? Suppose we have a platinum-iridium cube. What would be the length of each side for that cube to have a mass of 1 kilogram?—The answer is that each edge of the cube would be just a trifle less than 3.57 centimetres long (or 1.4 inches long in common American measure).

That makes the kilogram seem a surprisingly small unit of mass

—a metal cube that one can practically enfold in the hand—considering that a quart of milk is also very nearly a kilogram.

We will take up this seeming paradox in the next section, but, for now, let us begin moving down the ladder of mass.

STEP 2

0.316 kilograms ($10^{-0.5}$ kg)
3.16 hectograms ($10^{0.5}$ hg)

A kilogram is 1 000 grams and a hectogram is 100 grams; a kilogram is therefore equal to 10 hectograms, and, as indicated, 0.316 kilograms is equal to 3.16 hectograms.

The hectogram is almost never used as a unit, though it is permitted in the SI version of the metric system. Generally, if you want to avoid the decimal, you convert to grams and, instead of saying 0.316 kilograms, you say 316 grams.

A mass of 3.16 hectograms is equal to very nearly 0.7 pounds, or 11.15 ounces, in common American measure. In this case, however, we are talking about the "avoirdupois pound" and the "avoirdupois ounce," where "avoirdupois" is from old French words meaning "merchandise of weight"—the materials which pounds and ounces were used to measure.

In the days before the metric system, however, various places set their own standards of mass. There were therefore many different pounds and ounces, a situation which, if it existed today, would endlessly complicate world commerce and perhaps even throttle it. From this, the metric system saved us, and would save us even more efficiently but for the long-continued resistance of the English-speaking nations to it (all of whom, but the United States, have by now given in).

A pound and ounce other than the common avoirdupois varieties, still in use after a fashion, are the troy pound and troy ounce. Whereas the avoirdupois pound is equal to 16 avoirdupois ounces, the troy pound is equal to *12* troy ounces. (The word "troy" is not from the ancient city of Priam and Hector, but seems to be from the French town of Troyes, which is located about 150 kilometres southeast of Paris. In Troyes, during the Middle Ages, important fairs were held at which much buying and selling was done. The "troy" system of measurement was carefully watched and controlled so that people would not feel cheated and would keep coming to the fair. That system therefore became popular.)

Since the avoirdupois pound is 16 ounces and the troy pound is 12 ounces, one would be forgiven for thinking that the troy pound is equal to 3/4 of an avoirdupois pound. That, however, is an example of the pitfalls in nonmetric measurements, for that would only be true if the ounces were the same in mass in both cases—but they are not!

The troy ounce is about 10 percent more massive than the avoirdupois ounce. In fact, 1 troy ounce is equal to 1.097 avoirdupois ounces. That means that 10 troy ounces are equal to 10.97 avoirdupois ounces, or very nearly the Step-2 mass. In fact, 3.16 hectograms are equal to 10.16 troy ounces.

Since a troy pound is equal to 12 troy ounces, it is also equal to 13.164 avoirdupois ounces, or to 0.82275 (nearly five sixths) of an avoirdupois pound.

An avoirdupois pound is equal to 4.5359 hectograms, while a troy pound is equal to 3.7324 hectograms.

To be sure, avoirdupois pounds and ounces are used exclusively in measuring the mass of almost everything the American public buys. However, there are some things that ordinary people rarely buy that are measured in troy measure even today. Gold, for instance, has its mass described in troy measure. That is also true of gems and of the drugs sold by the apothecary (or druggist). In fact, troy measure is sometimes called "apothecary measure."

This explains the old conundrum: "Which weighs more, a pound of feathers or a pound of gold?"

The first impulse is to suppose that the gold is more massive because gold is so much "heavier" than feathers. Then we realize that there is a huge volume of feathers being referred to, and a small quantity of gold, for both have a mass of "a pound," so we answer triumphantly that a pound of feathers and a pound of gold are equal in mass.

And we are wrong! It is an avoirdupois pound of feathers and a troy pound of gold, so that a pound of feathers is substantially more massive than a pound of gold.

In the world of life, an adult Yorkshire terrier has been reported to be as low as 3.1 hectograms in mass, while a swan's egg is about 3.5 hectograms.

The largest diamond ever discovered is the "Cullinan" (named for the man who discovered the mine in South Africa, from which it was taken in 1905). Its original mass, as discovered, was 6.212 hectograms, or just twice the mass of the Yorkshire terrier mentioned in the previous paragraph.

STEP 3

0.1 kilograms (10^{-1} kg)
1 hectogram (10^0 hg)

A hectogram is equal to 3.532 avoirdupois ounces or 3.215 troy ounces. This means that we are near the quarter-pound level.

The American shopper usually buys butter in quarter-pound units, a square prism of that mass of butter wrapped in wax-paper. When you buy such a prism of butter, you are buying 1.134 hectograms.

For those of us who think of insects as tiny things, it is interesting (or, perhaps, horrifying) to remember that a few are not altogether tiny. The largest insect, mentioned earlier, is the Goliath beetle, and one of them was found to have a mass of almost exactly 1 hectogram.

The Cullinan diamond was cut into nine large stones and about one hundred small ones, all flawless. The largest diamond produced from the stone was "The Star of Africa," and it is the largest cut and polished diamond known. It is 1.06 hectograms in mass or just about the mass of a Goliath beetle.

STEP 4

0.031 6 kilograms ($10^{-1.5}$ kg)
3.16 dekagrams ($10^{1.5}$ dag)

A dekagram is equal to 10 grams, and at the Step-4 mass-level, we reach the range of ounces. A troy ounce is equal to 3.11 dekagrams, and an avoirdupois ounce is equal to 2.835 dekagrams.

A hen's egg of average size would have a mass of about 5.25 dekagrams (though jumbo eggs may reach a mass of 7 dekagrams).

The smallest primate is the pygmy marmoset, which can weigh, as an adult, as little as 4 dekagrams. Marmosets have brains that, considering their body size, are the largest known. A marmoset brain is equal to as much as one eighth the mass of its body (whereas a human brain is about one fiftieth the mass of its body). However, the absolute size of the marmoset brain is so small that the marmoset remains an unintellectual creature, though undoubtedly brighter than other animals its size.

The smallest living member of the group of animals usually referred to as carnivores is a particularly small species of weasel, where the adults have masses of as little as 3.5 dekagrams. Such a carnivore, meeting a Goliath beetle, would meet with an insect three times its own mass.

STEP 5

0.01 kilograms (10^{-2} kg)
1 dekagram (10^{0} dag)

A dekagram is 0.3527 avoirdupois ounces, or 0.3215 troy ounces, so we are dealing, roughly, with masses of a third of an ounce.

At this Step-5 mass-level, we are in the range of very small mammals indeed. Mice are the smallest mammals commonly met with, so that "as small as a mouse" has become a frequently repeated comparison. The common house mouse, the one most frequently seen, may have a mass of 2 dekagrams. The smallest species of mouse, however, is the Old World harvest mouse, which has a mass of a dekagram at most, though some adults are as little as half a dekagram in mass.

Among birds, the wren has a mass of about a dekagram, though there are some species with masses of just under half a dekagram.

STEP 6

0.003 16 kilograms ($10^{-2.5}$ kg)
3.16 grams ($10^{0.5}$ g)

The gram was originally thought of as the base-unit of mass, but it is too small to be convenient. There are about 28.35 grams to the avoirdupois ounce and 31.1 grams to the troy ounce.

The ounce is, in fact, inconveniently large to use as a comparison at this point. Although the ounce is the smallest unit of mass in everyday use in the United States, it is divided into "drams." (There is also a dram used as a measure of volume, described earlier in the book.)

There are actually two different drams used as measures of mass. The avoirdupois ounce is divided into 16 avoirdupois drams, so that each avoirdupois dram equals 1.77 grams. The troy ounce, on the other hand, is divided into 8 troy drams, so that each troy dram equals 3.89 grams. The Step-6 mass-level, then, stands nearly at the troy dram mark.

The troy dram, as you see, is equal to about 2.2 avoirdupois drams. Comparing the two drams, then, is exactly like comparing kilograms and avoirdupois pounds. Since the drams are usually referred to without the qualifying adjectives, there is plenty of ground for confusion.

A troy dram is divided into 2.5 pennyweights, so that a troy ounce is equal to 8 × 2.5, or 20 pennyweights. (The "pennyweight" is so-called because the English shilling, which was originally a troy ounce of silver, is divided into 20 pennies, or pence.) A pennyweight is equal to 1.5555 grams, and that, presumably, is what a silver penny, if there were such a thing, would weigh.

Apothecaries divide a troy dram into 3 "scruples" so that a scruple weighs 1.3 grams or five sixths of a pennyweight. (Is there anyone who doesn't see the superiority of the metric system after all this, by the way?)

At this mass-level, we have reached the smallest of all warm-blooded organisms. The smallest mammal is the pygmy shrew, which probably never attains a mass of more than 2.5 grams. The largest of them has a mass of less than a troy dram. The smallest of them, may, even as an adult, have a mass of no more than a pennyweight. The pygmy shrew has a mass only one quarter that of the smallest mouse.

The smallest bird is the bee-hummingbird, which is about the mass of a pygmy shrew; it is about a fifth the mass of the wren.

The largest butterfly known is a species from the Solomon Islands that can have a mass of up to 5 grams. The bee-hummingbird or the pygmy shrew has a mass less than half that of that butterfly, and only 1/50 that of a Goliath beetle.

STEP 7

0.001 kilograms (10^{-3} kg)
1 gram (10^{0} g)

The smallest land vertebrates, tiny lizards and newts, are probably no more than a gram, or a bit less, in mass. The fact that cold-blooded vertebrates, such as reptiles and amphibia, do not have to maintain constant, high body temperatures allows them to be less massive than warm-blooded vertebrates can be. (The smaller a life-form, the more rapidly it loses heat.)

STEP 8

0.000 316 kilograms ($10^{-3.5}$ kg)
3.16 decigrams ($10^{0.5}$ dg)

A decigram is a tenth of a gram, and although it is a permitted unit in the SI version, it, like dekagram and hectogram, is almost never used. Ordinarily, if people wish to deal with a mass that is equal to

0.316 grams, they do not say 3.16 decigrams, but 316 milligrams.

A nonmetric unit that is familiar to most people, even in lands that use the metric system, is the "carat," which is most often used to describe the mass of diamonds. (The word "carat" is from an old word for a small bean such as that of the carob tree.)

Originally, there were many different carats, all in the neighborhood of 2 decigrams. In 1913, the United States adopted a carat that was exactly 2 decigrams in mass (or 200 milligrams). This is called the "metric carat" and is now worldwide in use. Thus, The Star of Africa, the largest cut diamond, is 530.2 metric carats in mass, and the Cullinan, from which it was cut, was 3 106 metric carats when first found. Other precious gems, such as rubies, emeralds, and sapphires, also are measured in carats.

A version of the word, spelled with a "k," is used to express the purity of gold and of other precious metals. The karat, however, is not a unit of mass, but of proportion. Fourteen-karat gold is a gold alloy that is 14/24 gold and 10/24 something else, but such a mixture can have any mass at all.

The smallest eggs laid by any birds are, as you might expect, those laid by the bee-hummingbird. They are only about 5 decigrams in mass (2.5 carats, if you want to compare them to diamonds).

STEP 9

0.000 1 kilograms (10^{-4} *kg*)
1 decigram (10^{0} *dg*)

A decigram is half a carat.

The smallest unit of mass in common American usage is the "grain," the name of which comes from the fact that it originally signified the mass of a typical grain of wheat.

The grain began as a unit in the troy system. There are 20 grains to a scruple, and 24 grains to a pennyweight. Since there are 3 scruples to a troy dram and 2.5 pennyweights to a troy dram, there are, either way, 60 grains to a troy dram. Since there are 8 drams to a troy ounce and 12 troy ounces to a troy pound, there are 480 grains to a troy ounce and 5 760 grains to a troy pound.

As it happens, there are exactly 7 000 grains to an avoirdupois pound. That means there are 437.5 grains to the avoirdupois ounce, and just a little over 27 1/3 grains to the avoirdupois dram.

As far as the metric system is concerned, each grain is equal to 0.648 decigrams.

STEP 10

0.000 031 6 kilograms ($10^{-4.5}$ kg)
3.16 centigrams ($10^{0.5}$ cg)

A centigram is a tenth of a decigram, or a hundredth of a gram, and, once again, while permitted in the SI version of the metric system, this is a unit that is very rarely used.

Water oozing from a small aperture will form drops that will have a mass of about 6 centigrams each. A grain, as I pointed out in the section above, is equivalent to 0.648 decigrams and, therefore, to 6.48 centigrams. It follows that a drop of water is just under a grain in mass, and 16 drops of water will weigh about a gram.

STEP 11

0.000 01 kilograms (10^{-5} kg)
1 centigram (10^{0} cg)

A metric carat is divided into twenty "points." Since the carat is equal to 2 decigrams, which is, in turn, equal to 20 centigrams, the point is equal to 1 centigram (or 0.01 grams).

Thus, in 1969, Richard Burton bought a 69.42-carat diamond for Elizabeth Taylor. Its mass was therefore 69.42 × 20, or 1388.4 points. Since Burton paid $1 200 000 for the stone, its price was $864.30 a point.

There are certain large moths that lay eggs that are about 1.7 centigrams in mass, the most massive insect eggs known. (If those eggs were diamonds, valued at the rate Burton's diamond was, they'd be almost $1 500 apiece.)

STEP 12

0.000 003 16 kilograms ($10^{5.5}$ kg)
3.16 milligrams ($10^{0.5}$ mg)

The milligram is a tenth of a centigram, a hundredth of a decigram, and—what is most important—a thousandth of a gram. Just as one skips downward from the kilogram to the gram in the ordinary handling of the metric system, one skips down again from the gram to the milligram. Thus, one speaks of 350 milligrams or 35 milligrams rather than 3.5 decigrams or 3.5 centigrams, respectively.

In the same way, the usual conversion figures given are that a carat is equal to 200 milligrams (rather than 2 decigrams), and a point is equal to 10 milligrams (rather than to 1 centigram). A grain is equal to 64.8 milligrams.

At the Step-12 mass-level, we are approaching the smallest vertebrates. The smallest freshwater fish known are tiny gobies from the Philippine Islands, adult specimens of which may have masses as little as 4 milligrams.

STEP 13

0.000 001 kilograms (10^{-6} kg)
1 milligram (10^{0} mg)

There are seawater fish, also gobies, found off the Marshall Islands, that are the least massive of all vertebrates (though they are a trifle longer than the freshwater gobies of the Philippines). These marine gobies have a mass of as little as 2 milligrams.

STEP 14

0.000 000 316 kilograms ($10^{-6.5}$ kg)
316 micrograms ($10^{2.5}$ μg)

Once we move down beyond the milligram, there will be new prefixes only every three orders of magnitude. Beyond the milligram is the microgram, which is a thousandth of a milligram, and, therefore, a millionth of a gram.

Take a drop of water that has oozed and fallen from a small aperture and divide it into 200 equal droplets. Each of those droplets would have a mass of about 316 micrograms. As for 316 micrograms of Richard Burton's diamond, valued in proportion to the whole, that mass would be worth $27.30.

STEP 15

0.000 000 1 kilograms (10^{-7} kg)
100 micrograms (10^{2} μg)

The smallest spider, a species living in Australia, has a mass of about 100 micrograms. The water droplets that fill the air in a light drizzle may have a mass of no more than 100 micrograms each.

STEP 16

0.000 000 031 6 kilograms ($10^{-7.5}$ kg)
31.6 micrograms ($10^{1.5}$ μg)

The smallest segmented worm (of the group to which the common earthworm belongs) is less than half a millimetre in length and its mass is in the Step-16 mass-range.

STEP 17

0.000 000 01 kilograms (10^{-8} kg)
10 micrograms (10^{1} μg)

The smallest crustaceans (of the group to which crabs and lobsters belong) are tiny water fleas that have masses of perhaps 8 micrograms.

STEP 18

0.000 000 003 16 kilograms ($10^{-8.5}$ kg)
3.16 micrograms ($10^{0.5}$ μg)

The smallest insects are tiny parasitic wasps with a wingspread of no more than a millimetre and with masses that may be as low as 5 micrograms. Even such small organisms are multicellular and are made up of a number of cells (though a very small number in comparison with larger organisms).

Still, there are unicellular organisms (those made up of a single cell) that approach the mass of such tiny multicellular organisms. The paramecium and the amoeba, best known to youngsters in their first school adventures with a microscope, have masses of perhaps 4 micrograms.

STEP 19

0.000 000 001 kilograms (10^{-9} kg)
1 microgram (10^{0} μg)

The smallest multicellular organisms are the rotifers, each of which is made up of a fixed (and small) number of cells. The smallest of these may have a mass of no more than 1 microgram, and from now on we must deal with nothing more massive than individual cells.

The human ovum, or egg cell, for instance, would have a mass of about 1.5 micrograms. This is very small compared to the eggs of birds—with which we are most familiar—but in the case of the human ovum, no food supply is needed except for the short interval before the egg is implanted in the wall of the uterus and begins to be nourished from the mother's bloodstream by way of the placenta. Other mammalian egg cells are equally small.

Nevertheless, even that small quantity of food suffices to make the ovum the largest cell in the human body of either sex. (The ovum is found only in the female, of course.)

STEP 20

0.000 000 000 316 kilograms ($10^{-9.5}$ kg)
316 nanograms ($10^{2.5}$ ng)

A nanogram is a thousandth of a microgram and, therefore, a billionth of a gram.

Mammalian ova are not the smallest eggs there are. The smallest insects are not very much larger than a mammalian egg cell, and so, of necessity, their eggs are considerably smaller. The tiniest insects lay eggs with a mass of about 200 nanograms, one seventh the mass of the human ovum, but then remember that that tiny egg must contain all the food the developing insect will have till it is sufficiently large to find food on its own. The egg, by the way, is about one twenty-fifth the mass of the insect that lays it.

STEP 21

0.000 000 000 1 kilograms (10^{-10} kg)
100 nanograms (10^{2} ng)

We are down in the range of fine dust. The typical particle of volcanic dust may have a mass of about 100 nanograms.

Such dust is smaller than any animal eggs, but plants can do better still. Certain orchids produce seeds with a mass of 80 nanograms each.

STEP 22

0.000 000 000 031 6 kilograms ($10^{-10.5}$ kg)
31.6 nanograms ($10^{1.5}$ ng)

At this mass-level, we have finally left eggs and seeds behind and are approaching the realm of the ordinary cells that make up the tissues of multicellular animals.

STEP 23

0.000 000 000 01 kilograms (10^{-11} *kg*)
10 nanograms (10^{1} *ng*)

The largest cells in human tissues have masses of up to about 15 nanograms, about 1/100 the mass of the human ovum. Human cells of average size, such as the busy cells of that chemical factory, the liver, have masses of half that size, about 8 nanograms.

STEP 24

0.000 000 000 003 16 kilograms ($10^{-11.5}$ *kg*)
3.16 nanograms ($10^{0.5}$ *ng*)

We are now in the range of tissue cells of less than average size.

STEP 25

0.000 000 000 001 kilograms (10^{-12} *kg*)
1 nanogram (10^{0} *ng*)

Insoluble materials may remain in suspension in water indefinitely, if they are divided into very small particles. If they are, the presence of tiny electric charges of similar sign on each particle (so that the particles repel each other), together with the hustling of the moving water molecules, suffices to keep the particles suspended. This is a "colloidal suspension," so-called because ordinary glues are examples of such suspensions and the Greek word for glue is "kolla." Thus, particles of very finely divided gold powder, with diameters of about half a micrometre, will form a permanent suspension. The masses of the individual particles are about 1 nanogram.

STEP 26

0.000 000 000 000 316 kilograms ($10^{-12.5}$ *kg*)
316 picograms ($10^{2.5}$ *pg*)

We have now worked our way down through the smallest of the ordinary tissue cells of multicellular organisms. Nevertheless, regions of life remain at further levels.

STEP 27

0.000 000 000 000 1 kilograms (10^{-13} kg)
100 picograms (10^2 pg)

Familiar "cells" that are not really ordinary tissue cells are the red blood corpuscles that pick up oxygen in the lung membranes and give it up to the tissue cells. They are not truly cells for they lack a nucleus, which is essential for cell division and reproduction. Red blood corpuscles arise through the division of true cells, but, in the process, they lose their nuclei and become mere transport mechanisms that wear out and disintegrate rather quickly, leaving no descendants. They are replaced by entirely new corpuscles produced by the nucleated cells whose function it is to keep them coming.

It is not surprising, then, that the red blood corpuscles are smaller than true cells. They have a mass of about 90 picograms apiece, are less than 1/80 the mass of liver cells, and less than 1/160 the mass of the largest body cells.

STEP 28

0.000 000 000 000 031 6 kilograms ($10^{-13.5}$ kg)
31.6 picograms ($10^{1.5}$ pg)

At the Step-28 mass-level, we can fairly talk about the internal "organelles" of tissue cells, the smaller structures of which cells are made up. Of these, the chloroplasts of plant cells are quite large as organelles go. They contain not only chlorophyll, which is essential to the series of chemical reactions that make use of sunlight in building complex compounds out of simple ones, but all the chemical and structural machinery required for those reactions. The chloroplasts in spinach leaves have a mass of about 60 picograms, two thirds the mass of a red blood corpuscle.

All the tissue cells of plants and animals, as well as most one-celled organisms, have nuclei, within which the reproductive machinery of the cell is contained. These generally run to half the mass of the chloroplasts.

That brings us to a very specialized cell, the spermatozoon, or sperm cell—which is the male analogue of the egg cell. Whereas the egg cell has a nucleus containing half the nuclear material of the ordinary cell of the female's tissue (a "half-nucleus," we might call it) surrounded by a comparatively vast supply of food, the sperm cell contains a half-nucleus only, attached to a lashing tail. It has virtually no food supply included, for all it needs to do is to swim to the egg cell and enter (in competition with millions of others, all trying for the same goal, though only *one* can succeed). It then adds its half-nucleus to the half-nucleus in the egg cell, producing a whole nucleus, and shares in exploiting the food supply.

The human sperm cell has a mass of about 17 picograms, less than 1/6 of the red blood corpuscle and about 1/80 000th that of the human egg cell.

STEP 29

0.000 000 000 000 01 kilograms (10^{-14} *kg*)
10 picograms (10^{1} *pg*)

There are some objects associated with multicellular organisms that can still be referred to. The human blood stream (and that of other vertebrates) contains certain structures called "platelets." They are not cells but are simply tiny structures that contain substances that bring about the clotting of blood. Any damage disrupts the platelets and leads to the formation of a blood clot. Individual platelets have a mass of perhaps 10 picograms, about three fifths that of a sperm cell and one ninth of a red blood corpuscle.

Just as the largest protozoa are more massive than the largest tissue cells of multicellular organisms, the smallest protozoa are less massive than the smallest tissue cells of multicellular organisms. The smallest protozoa may have a mass of no more than 7 picograms. This is perhaps a millionth the mass of the largest protozoa, and about a thousandth the mass of an average tissue cell.

Below the full range of protozoan cells, however, are a group of still smaller organisms called "bacteria." These are neither plant nor animal cells but are more primitive than either, lacking a nucleus. Unlike the case of red blood corpuscles, this does not mean bacteria do not reproduce. They do so with great facility, but the DNA that makes that possible is not organized within a well-defined nucleus, but is distributed throughout the cell.

Such nonnucleated ("prokaryotic") cells are more primitive than those with nuclei ("eukaryotic"), and the cell only works, apparently, under such nonefficient nuclear conditions, if it is very small. (There are also prokaryotic cells with chlorophyll, and these are called "blue-green algae.") Even the largest prokaryotic cells have a mass of no more than 7 picograms, less than a thousandth the mass of a typical eukaryotic cell.

STEP 30

0.000 000 000 000 003 16 kilograms ($10^{-14.5}$ *kg*)
3.16 picograms ($10^{0.5}$ *pg*)

Just as plant seeds can be smaller than any animal ovum, so the plant analogue of sperm cells (pollen grains) can be smaller than any animal sperm cells. Some grasses produce pollen grains that are no more than 5 picograms in mass, less than one third the mass of a sperm cell.

STEP 31

0.000 000 000 000 001 kilograms (10^{-15} *kg*)
1 picogram (10^{0} *pg*)

We are now well within the range of bacterial size. The human large intestine is loaded with small bacteria that help decompose the food, produce some vitamins which we absorb, and, in general, live with us in peace and ordinarily do us no harm. These are *E. coli* cells, and the individual *E. coli* cell is perhaps 2 picograms in mass.

In eukaryotic cells, there are numerous small "mitochondria" which contain the apparatus for breaking down molecules in such a way as to liberate, trap, and put to constructive use some of the chemical energy contained in those molecules, making use of oxygen for the purpose. The mitochondria of human liver cells are about 1.5 picograms in mass. This is about 1/40 the mass of a typical chloroplast (which shows, in a way, the greater difficulty of building up complex molecules, than tearing them down).

STEP 32

0.000 000 000 000 000 316 kilograms ($10^{-15.5}$ *kg*)
316 femtograms ($10^{2.5}$ *fg*)

A femtogram is a thousandth of a picogram and a trillionth of a gram. At the Step-32 mass-level, we are in the range of the smaller bacteria.

STEP 33

0.000 000 000 000 000 1 kilograms (10^{-16} *kg*)
100 femtograms (10^{2} *fg*)

At the Step-33 level, we are reaching the stage where it is difficult to pack into cells all the machinery needed for life. A bacterial cell with a mass of 100 femtograms can indeed contain the various substances needed for life, but only in minimal quantities, and with little space left for reserve.

This does not, however, mean that we are reaching the lower limit of life. It is possible for living things to lack some of the essentials and thus be unable to live on their own, and yet to penetrate other, more complete, cells, and to multiply there, making use of the host cell's chemical machinery for their own ends.

The largest of such parasitic "subcells" are the rickettsia. Some of these are responsible for well-known human diseases. The rickettsial organism that causes typhus fever, for instance, has a mass of about 54 femtograms.

STEP 34

0.000 000 000 000 000 031 6 kilograms ($10^{-16.5}$ kg)
31.6 femtograms ($10^{1.5}$ fg)

The smallest bacteria now known are "pleuropneumonia-like organisms" (PPLO), which were first discovered in sewage. These are the smallest cells that contain within themselves whatever chemical systems are needed for independent, nonparasitic life. These can have masses as small as 20 femtograms, so that the smallest free-living cells are less than half the mass of the rickettsia.

STEP 35

0.000 000 000 000 000 01 kilograms (10^{-17} kg)
10 femtograms (10^{1} fg)

Less massive than the rickettsia are a range of organisms called "viruses," all of which parasitize cells, and a few of which are responsible for some of the commonest human diseases.

One of the largest viruses is that which causes cowpox, and which is used for immunization against smallpox (a process called "vaccination," from the Latin word for "cow"). The cowpox virus may have a mass of only 5.6 femtograms.

STEP 36

0.000 000 000 000 000 003 16 kilograms ($10^{-17.5}$ kg)
3.16 femtograms ($10^{0.5}$ fg)

We are now sinking through the range of moderate-sized viruses.

STEP 37

0.000 000 000 000 000 001 kilograms (10^{-18} kg)
1 femtogram (10^0 fg)

The virus that causes influenza is not far below the 1 femtogram mark. Its mass may be 0.8 femtograms.

STEP 38

0.000 000 000 000 000 000 316 kilograms ($10^{-18.5}$ kg)
316 attograms ($10^{2.5}$ ag)

The prefix "atto-" (AT-uh), symbolized "a," represents a quintillionth of a basic measure. An attogram is a quintillionth of a gram and is symbolized "ag." The prefix is from a Danish word "atten" meaning "eighteen" since an attogram is 10^{-18} grams.

A particular group of viruses that infects bacterial cells are called "bacteriophages." Large members of this group have masses of up to 500 attograms.

STEP 39

0.000 000 000 000 000 000 1 kilograms (10^{-19} kg)
100 attograms (10^2 ag)

The first virus ever to be discovered was one which causes "tobacco mosaic disease," and it is, naturally, the "tobacco mosaic virus." It has a mass of about 70 attograms.

STEP 40

0.000 000 000 000 000 000 031 6 kilograms ($10^{-19.5}$ kg)
31,6 attograms ($10^{1.5}$ ag)

The chromosomes found in cell nuclei are divided into "genes," each of which is thought to govern the formation of a particular enzyme controlling a particular chemical reaction. The genes are similar in chemical nature, and in size, to the viruses. In fact, viruses probably control the cells they infest by themselves superseding the functions of the cell's own genes. A gene may have a mass of 40 attograms.

STEP 41

0.000 000 000 000 000 000 01 kilograms (10^{-20} *kg*)
10 attograms (10^{1} *ag*)

We are dealing now with viruses well below average in size. A bacteriophage which infests *E. coli* has a mass of about 10 attograms, while the virus which causes yellow fever has a mass of about 5.6 attograms.

STEP 42

0.000 000 000 000 000 000 003 16 kilograms ($10^{-20.5}$ *kg*)
3.16 attograms ($10^{0.5}$ *ag*)

There are, in cells, particularly small organelles called "ribosomes," which are the sites where protein molecules are synthesized. A ribosome of *E. coli* has a mass of about 4.7 attograms.

STEP 43

0.000 000 000 000 000 000 001 kilograms (10^{-21} *kg*)
1 attogram (10^{0} *ag*)

Now, at last, we are approaching the limits of life. Among the very smallest viruses is that which causes hoof-and-mouth disease in cattle. This virus has a mass of perhaps as little as 0.7 attograms.

Between the hoof-and-mouth virus and the giant sequoia tree there is a stretch of 27 orders of magnitude. This is like comparing the mass of a human being to that of a small star, and that is an impressive indication of the mass-range over which life can exist.

In the realm of nonlife, the tiny particles in colloidal suspensions show a number of their characteristic properties through the fact that those tiny particles tend to scatter light waves. As the particles grow smaller, however, a point is reached where light waves, tiny though they be, step over those particles, so to speak, and are therefore not scattered. Particles with a mass of 1 attogram or less are too small to form colloidal suspensions, therefore. At this point we begin to encounter true solutions.

Though we sink below the level of life and of colloids, there remain objects to consider. The simplest forms of life are little more than a handful of giant molecules (collections of atoms, bound quite firmly together) of two types chiefly: nucleic acids and proteins. Where the properties we associate with life are not required, such molecules can be less massive than the least massive life form.

Of course, there is overlapping, as in almost all divisions, so that there are nonliving protein molecules that are larger than the smallest viruses. For instance, the characteristic protein of muscle is "myosin," and there are some molecules of myosin that have masses of 1.03 attograms, nearly half again as massive as the hoof-and-mouth virus.

STEP 44

0.000 000 000 000 000 000 000 316 kilograms ($10^{-21.5}$ *kg*)
0.316 attograms ($10^{-0.5}$ *ag*)
190 000 daltons ($10^{5.28}$ *D*)

In the SI version of the metric system, there is no prefix representing a smaller division than "atto-." We will therefore remain with that prefix to the end.

We will, however, introduce a non-SI unit for the sake of convenience. Chemists give the mass of molecules in terms of something traditionally called "molecular weight." The unit of "molecular weight" is the "atomic mass unit," which is just about the mass of the least massive atom. The atomic mass unit is also called the "dalton" after the name of the English chemist who first developed the modern atomic theory. We will use the "dalton" as the unit, and symbolize it as "D." There are just about 600 000 daltons in an attogram, so that the 0.316 attograms of the Step-44 mass-level are equivalent to 190 000 daltons.

Some of the most important proteins are "enzymes," which control the speed of chemical reactions and, in that way, by their presence in particular proportions, control the chemical machinery of the body down to the finest details. One of the larger enzyme molecules is "glutamic acid dehydrogenase," an enzyme which controls the rate at which a molecule known as glutamic acid (much smaller than the enzyme) loses two hydrogen atoms. It has a mass of 0.416 attograms (250 000 daltons).

As another example, there are gamma globulins, members of a class of proteins in the bloodstream that is extremely important in maintaining the body's immunity to disease. These come in a wide range of size, but a typical gamma globulin would have a molecular weight of 0.266 attograms (160 000 daltons).

STEP 45

0.000 000 000 000 000 000 000 1 kilograms (10^{-22} *kg)*
0.1 attograms (10^{-1} *ag)*
60 000 daltons ($10^{4.78}$ *D)*

Hemoglobin is one of the best-known proteins. It is the one which gives blood its red color, and which takes up oxygen in the lungs. (Actually, hemoglobin is red only after it takes up oxygen and becomes "oxyhemoglobin," but when we bleed, the blood that issues from the wound, if it is not oxygenated to begin with, takes up oxygen at once, so that we always see it as red.)

A molecule of hemoglobin has a mass of about 0.11 attograms, or 68 000 daltons.

STEP 46

0.000 000 000 000 000 000 000 031 6 kilograms ($10^{-22.5}$ *kg)*
0.031 6 attograms ($10^{-1.5}$ *ag)*
19 000 daltons ($10^{4.28}$ *D)*

Hemoglobin consists of four portions, fitted neatly together. In muscle, there is an oxygen-gathering protein rather resembling a molecule of hemoglobin, made up of only a single one of those portions. It seems to be a "quarter-hemoglobin" molecule and it is called "myoglobin." A molecule of myoglobin has a mass of about 0.028 attograms (17 000 daltons).

The pancreas secretes a fluid into the small intestine, one that contains a variety of digestive enzymes that help break up the complex molecules of food into smaller fragments that can be absorbed. One of these enzymes, which helps break up protein molecules, is called "trypsin." A molecule of trypsin has a mass of about 0.04 attograms (23 800 daltons).

STEP 47

0.000 000 000 000 000 000 000 01 kilograms (10^{-23} *kg)*
0.01 attograms (10^{-2} *ag)*
6 000 daltons ($10^{3.78}$ *D)*

We are now in the range of the smallest protein molecules. The hormone insulin is a protein that is essential to the proper functioning of those body reactions which break down glucose (the form of sugar that acts as the body's immediate fuel) and extract energy from it. Insulin is formed in the pancreas, and when the pancreas fails in this task, the result is "diabetes." The mass of a molecule of insulin is very nearly 0.01 attograms (5 800 daltons).

STEP 48

0.000 000 000 000 000 000 000 003 16 kilograms ($10^{-23.5}$ kg)
0.003 16 attograms ($10^{-2.5}$ ag)
1 900 daltons ($10^{3.28}$ D)

Proteins and nucleic acids are called "organic molecules," because they are characteristic of living (or once-living) organisms. The definition has been generalized so that organic molecules are now considered those which, like those in organisms, consist of chains or rings of carbon atoms to which other types of atoms are attached—even if the particular molecule never appears in living organisms.

Because carbon atoms have the ability to attach to each other in long chains, branched or unbranched, and in complicated ring systems, organic molecules can be much larger and more massive than other molecules (inorganic) are.

Nevertheless, there are inorganic molecules that are respectable in size. Ammonium phosphotungstate, for instance, has a molecule containing 77 atoms, not one of which is carbon. The mass of a molecule of ammonium phosphotungstate is about 0.005 attograms (2 986 daltons), just above the Step-48 mass-level. A related molecule, ammonium phosphomolybdate, has a mass of 0.003 2 attograms (1 931 daltons).

Compare this with "oxytocin," a small hormone molecule produced by the pituitary gland. A molecule of oxytocin has a mass of only 0.001 7 attograms (1 007 daltons).

STEP 49

0.000 000 000 000 000 000 000 001 kilograms (10^{-24} kg)
0.001 attograms (10^{-3} ag)
600 daltons ($10^{2.78}$ D)

We have now reached molecules of only moderate size, with many representatives, both organic and inorganic.

There are, for instance, organic molecules smaller than enzymes, without which enzymes could not carry out their functions. These are "coenzymes" and are, so to speak, the cutting edge of the enzyme. One example is "diphosphopyridine nucleotide," the molecules of which contain 71 atoms, of which 21 are carbon. The mass of one of its molecules is 0.001 1 attograms (663.4 daltons).

Among the inorganic molecules, to pick one almost at random, hydrated samarium bromate is made up of molecules containing 40 atoms. One of these molecules has a mass of 0.001 16 attograms (696.2 daltons).

You might wonder why a molecule with 71 atoms should have a smaller mass than one with 40 atoms, but, as will soon be explained, atoms of different kinds have different masses to begin with.

STEP 50

0.000 000 000 000 000 000 000 000 316 kilograms ($10^{-24.5}$ *kg*)
0.000 316 attograms ($10^{-3.5}$ *ag*)
190 daltons ($10^{2.28}$ *D*)

Protein molecules are made up of chains of small molecules called "amino acids." A typical protein molecule would contain hundreds, or even thousands, of these in the chain. Similarly, nucleic acids are made up of chains of smaller molecules called "nucleotides." The huge molecules of starch or cellulose are made up of chains of smaller molecules called "glucose."

Of these small molecules, the nucleotides are the most complex. A typical one is adenylic acid, which, with 37 atoms, has a molecular mass of 0.000 577 attograms (347.2 daltons). Of the amino acids, the most complex is tryptophan, which has a 27-atom molecule and a molecular mass of 0.000 340 attograms (204.2 daltons). As for glucose, with its 24-atom molecule, its molecular mass is 0.000 300 (180.2 daltons).

Among inorganic molecules, basic copper carbonate is a common copper ore, which, as a mineral, is called "malachite." It has a ten-atom molecule, and a molecular mass of 0.000 368 (221.1 daltons).

I have mentioned the fact that different atoms have different masses, and, as might be expected, the most massive atoms are more massive than the least massive molecules.

Thus, the most massive atoms known have masses of more than 0.000 415 attograms (250 daltons). These atoms have, however, been observed only in the laboratory, where they have been manufactured in trace amounts from smaller atoms. What is more, these massive atoms are very unstable and break down, relatively quickly, to smaller atoms.

The most massive of the atoms that occur on Earth in significant quantities are those of uranium, which are radioactive but which break down only after a long period of time—so long that sizable quantities of uranium that existed at the time of Earth's formation still exist today. The more common of the two uranium isotopes has an atomic mass of 0.000 396 attograms (238 daltons).

The most massive atom that is quite stable is that of bismuth, of which there is only one isotope that occurs in substantial quantities on Earth. Its atomic mass is 0.000 348 attograms (209 daltons).

Gold has an atomic mass of 0.000 328 attograms (197 daltons), and the mass of an atom of one particular isotope of iridium is 0.000 316 attograms (191 daltons).

STEP 51

0.000 000 000 000 000 000 000 000 1 kilograms (10^{25} *kg*)
0.000 1 attograms (10^{-4} *ag*)
60 daltons ($10^{1.78}$ *D*)

We have now reached the smallest of the amino acids, which is glycine. This has a ten-atom molecule with a mass of 0.000 125 attograms (75 daltons).

Urea is a breakdown product of proteins and is excreted in the urine (from which it derives its name). Its molecule is made up of eight atoms (four hydrogens, two nitrogens, a carbon, and an oxygen), and each of these molecules has a mass of 0.000 1 attograms (60 daltons).

Acetic acid, which gives vinegar its sour taste, has molecules made up of eight atoms (four hydrogens, two carbons, and two oxygens), and these atoms have just the mass of those of urea.

These are quite simple organic molecules, and at this mass-level, inorganic molecules (often made up of more massive atoms than organic molecules are) are simpler still. Thus, the molecule of sodium chloride (common table salt) consists of two atoms, a sodium atom and a chlorine atom, and its mass is 0.000 097 attograms (58.4 daltons).

As far as individual atoms are concerned, we are down into the lower-middle range. The atoms of the three similar metals—iron, cobalt, and nickel—have masses that range from 0.000 09 attograms (54 daltons) to 0.000 106 attograms (64 daltons).

STEP 52

0.000 000 000 000 000 000 000 000 031 6 kilograms ($10^{-25.5}$ *kg*)
0.000 031 6 attograms ($10^{-4.5}$ *ag*)
19 daltons ($10^{1.28}$ *D*)

At the Step-52 mass-level, molecules are very simple indeed. The simplest of all the organic molecules are those of methane, these being made up of one carbon atom and four hydrogen atoms. Each of these molecules has a mass of 0.000 026 6 attograms (16 daltons).

At this stage, by the way, the mass of a molecule changes significantly with the nature of the isotope that is found in the molecule. Both carbon and hydrogen are made up of two varieties, the more massive of which is quite uncommon. In the case of hydrogen, the larger has a mass of 2 daltons, rather than 1, and in the case of carbon, 13 daltons rather than 12. If the more massive isotope is included in all five atoms of methane, the total mass of the molecule is 0.000 034 9 attograms (21 daltons), which is nearly one third more massive than the ordinary molecule. However, since only one out of 6 700 hydrogen atoms is the more massive isotope, and only 1 out of 90 carbon atoms is, only 1 out of 44 000 000 000 methane molecules would have only the more massive varieties and would achieve the maximum stable mass. (There are still more massive isotopes in the case of both hydrogen and carbon, but these are radioactive and not stable.)

Among the inorganic molecules, the water molecules consist of one oxygen atom and two hydrogen atoms, and each of these molecules has a mass of 0.000 035 5 attograms (18 daltons).

As for the atoms themselves, we are now dealing with the simplest. Each atom of the gas fluorine, the ninth simplest atom, has a mass of of 0.000 037 5 attograms (19 daltons).

Each atom of carbon, the sixth simplest atom, has a mass of 0.000 02 attograms (12.01 daltons), but this is an average value. As I said just above, carbon atoms come in two stable isotopes. The less massive has a mass of 0.000 019 95 attograms (*exactly* 12 daltons). Indeed, that is how one defines a dalton; it is just one twelfth the mass of the less massive isotope of carbon (carbon-12).

STEP 53

0.000 000 000 000 000 000 000 000 01 kilograms (10^{-26} kg)
0.000 01 attograms (10^{-5} ag)
6 daltons ($10^{0.78}$ D)

There are scarcely any molecules remaining at this level. Lithium hydride has molecules made up of two atoms each: one of lithium, the third simplest element, and one of hydrogen, the very simplest. Each molecule has a mass of 0.000 013 2 attograms (8 daltons).

Lithium itself comes in three stable isotopes, the smallest having a mass of 0.000 008 3 attograms (5 daltons).

STEP 54

0.000 000 000 000 000 000 000 000 003 16 kilograms ($10^{-26.5}$ kg)
0.000 003 16 attograms ($10^{-5.5}$ ag)
1.9 daltons ($10^{0.28}$ D)

Now we are at the bottom of the ladder as far as atoms are concerned. The second simplest element is helium, the atoms of which combine with no other atom, not even with each other. The helium atoms exist in two stable isotopes. The larger, and by far the more common, has a mass of 0.000 006 65 attograms (4 daltons), while the smaller has one of 0.000 005 attograms (3 daltons).

The smallest atoms are those of hydrogen, which exist as two stable isotopes. The larger (deuterium) and by far the less common has a mass of 0.000 003 32 attograms (2 daltons), while the smaller and far more common isotope has a mass of 0.000 001 66 attograms (1 dalton). The smallest atom, in other words, has a mass of slightly more than a millionth of an attogram.

Atoms are made up of particles still smaller in physical dimensions (as far as "physical dimensions" has meaning at that level of material existence), and these are called "subatomic particles."

There are subatomic particles that are not viewed as part of ordinary atoms under ordinary circumstances, but that can be produced in the laboratory in very small quantities. Some of these are quite massive for such objects, more massive in fact than the smallest atoms (another case of overlap). Some subatomic particles have masses in the neighborhood of 2 daltons and are as massive as the deuterium atom.

(There is even a subatomic particle, called the "magnetic monopole," that has never yet been detected but that, for theoretical reasons, is thought to be 10 000 000 000 000 000 daltons in mass—or more massive than an amoeba.)

Most of the mass of an atom is to be found in the atomic nucleus, which is composed of two types of particles: protons and neutrons. These are sometimes lumped together as "nucleons" because they occur in the nucleus. Protons and neutrons, together with all subatomic particles that are more massive still, are called "baryons," from a Greek word for "heavy."

The neutron has a mass of 0.000 001 674 8 attograms (1.0087 daltons), and it breaks down with moderate speed to form a proton. The proton is slightly smaller than the neutron, since it has a mass of 0.000 001 672 5 attograms (1.0073 daltons). It seems to be a rule that baryons can only break down into other, less massive baryons; and since the proton is the least massive baryon, it cannot break down, and is stable. (It is possible, however, according to some current theories, that protons break down into less massive nonbaryons with extreme slowness. They do so with such slowness that such breakdowns have never been detected, though experiments are now in progress to attempt the detection of such breakdowns if they exist.)

In continuing the descent down the ladder of mass, we will now abandon the dalton. Albert Einstein, in his theory of relativity, suggested that mass and energy were interconvertible, and experiment found him to be completely correct. Units of energy can, therefore, be used to express quantities of mass. A very small quantity of mass is equivalent to a very large amount of energy, so that it is easy to find a useful unit of energy so small as to have convenient meaning for masses less than that of an atom.

There is, for instance, the "electron volt" (symbolized "ev"), which is not part of the SI system, but which is frequently used by nuclear physicists. It is a very small unit of energy indeed, so small that 1 dalton is equal to 931 500 000 electron volts.

STEP 55

0.000 000 000 000 000 000 000 000 001 kilograms (10^{-27} kg)
0.000 001 attograms (10^{-6} ag)
560 000 000 electron volts ($10^{8.75}$ ev)

Less massive than even the least massive baryon are the members of a group of subatomic particles of intermediate size. These are called "mesons," from the Greek word for "intermediate." None of them are stable. The most massive mesons reach the Step-55 mass-level. The eta-meson, for instance, has a mass of 0.000 000 98 attograms, or 548 800 000 electron volts. It has not quite three fifths the mass of a proton.

STEP 56

0.000 000 000 000 000 000 000 000 000 316 kilograms ($10^{-27.5}$ kg)
0.000 000 316 attograms ($10^{-6.5}$ ag)
177 000 000 electron volts ($10^{8.25}$ ev)

This brings us to the lower limit of the meson range. The least massive of the mesons is the pi-meson (or pion), which has a mass of 0.000 000 245 attograms, or 137 000 000 electron volts. The pion is a little over one seventh as massive as a proton.

The mesons and baryons are capable of interacting with each

other very rapidly and are therefore said to be subject to the "strong interaction." They are, collectively, called "hadrons." Subatomic particles that are (mostly) still less massive than the mesons can interact only comparatively slowly, and are subject to the "weak interaction." Such less active particles are therefore called "leptons" from a Greek word for "weak."

The most massive leptons known have masses well up in the meson range, and possibly even in the baryon range, but all such massive leptons are very unstable.

The most common of the massive leptons, for instance, is the "muon," which has a mass of 0.000 000 188 attograms, or 105 800 000 electron volts. It is three fourths as massive as a pion, and one ninth the mass of a proton.

STEP 57

0.000 000 000 000 000 000 000 000 000 1 kilograms (10^{-28} kg)
0.000 000 1 attograms (10^{-7} ag)
56 000 000 electron volts ($10^{7.75}$ ev)

We are now well down into the lepton range. Less massive than the muon by a good deal is the electron, which is far below the Step-57 mass-level. Indeed, there is no particular particle to which we can point as an example of this stage in the descent. We can only deal with groups of electrons.

For instance, each atom of the element uranium contains 92 electrons. These electrons, taken together, have a mass of just under 0.000 000 084 attograms (or, about 47 000 000 electron volts).

STEP 58

0.000 000 000 000 000 000 000 000 000 031 6 kilograms ($10^{-28.5}$ kg)
0.000 000 031 6 attograms ($10^{-7.5}$ ag)
17 700 000 electron volts ($10^{7.25}$ ev)

Each atom of the element bromine contains 35 electrons. These electrons, taken together, have a mass of 0.000 000 031 9 attograms, or about 17 900 000 electron volts.

STEP 59

0.000 000 000 000 000 000 000 000 000 01 kilograms (10^{-29} kg)
0.000 000 01 attograms (10^{-8} ag)
5 600 000 electron volts ($10^{6.75}$ ev)

Each atom of the element sodium has 11 electrons. These electrons, taken together, have a mass of just about 0.000 000 01 attograms, or about 5 600 000 electron volts.

STEP 60

0.000 000 000 000 000 000 000 000 000 003 16 kilograms ($10^{-29.5}$ kg)
0.000 000 003 16 attograms ($10^{-8.5}$ ag)
1 770 000 electron volts ($10^{6.25}$ ev)

Each atom of the element beryllium has 4 electrons. These electrons, taken together, have a mass of 0.000 000 003 65 attograms, or 2 044 000 electron volts.

STEP 61

0.000 000 000 000 000 000 000 000 000 001 kilograms (10^{-30} kg)
0.000 000 001 attograms (10^{-9} ag)
560 000 electron volts ($10^{5.75}$ ev)

Now, at last, having reached a billionth of an attogram (a quadrillionth of a quadrillionth of a kilogram), we have reached very nearly to the level of a single electron. An electron has a mass of 0.000 000 000 91 attograms, or 511 000 electron volts. It has a mass 1/1 837 that of a proton.

The electron is the smallest particle known that has a definite nonzero "rest mass." That is, the electron has a mass of 511 000 electron volts even when it is motionless with respect to its surroundings. When it moves, it possesses also energy of motion (kinetic energy), which, by Einstein's theory, is equivalent to additional mass. This additional mass is (at ordinary speeds, at least) very small compared with the rest mass.

There is also a positron, with a rest mass of 511 000 electron volts, but a positron is merely an electron that carries a positive charge rather than a negative one. The two particles may be considered as mirror images, so to speak.

Until recently, it was suspected that the electron and positron were the smallest particles-with-mass that could *possibly* exist. To be sure, there are leptons that are smaller than the electron and positron. There are various "neutrinos" and "antineutrinos" (at least three of each are known), but these are supposed to be without mass. That is, they have zero rest mass.

All objects with zero rest mass move through a vacuum at precisely the speed of light, so that neutrinos and antineutrinos would have masses equal to the equivalent of their kinetic energy, but that would be small indeed.

In 1980, however, experiments seemed to indicate that neutrinos may, indeed, have a very tiny rest mass of about a ten-thousandth that of an electron. (Tiny though that might be, it would be much greater than the mass-equivalence of their kinetic energy.) These experiments are rather shaky and may prove to be wrong. Let us suppose they are right, however. If so, we will have to skip eight steps to get down to the neutrino mass-level—

. . .

STEP 69

0.000 000 000 000 000 000 000 000 000 000 000 1 kilograms (10^{-34} kg)
0.000 000 000 000 1 attograms (10^{-13} ag)
56 electron volts ($10^{1.75}$ ev)

The rest mass of the neutrino would be somewhere near this Step-69 mass-level (a ten-trillionth of an attogram) if the experimental determinations hold up.

Even if the neutrinos turn out to have rest mass, however, there still remain photons. These seem, in even the most rigorous experiments, to have truly zero rest mass. They must move at the speed of light, therefore (in fact, light itself consists of photons), and this gives them kinetic energy that works out to the equivalent of a tiny bit of mass.

The amount of energy present in photons depends upon their wavelength. The shorter the wavelength (that is, the higher the frequency), the greater the kinetic energy present, and the larger the mass-equivalence.

Very short-wave photons have a mass-equivalence equal to some of the lighter particles. The photons of X rays have a mass-equivalence equal to the rest mass of electrons, for instance. If we want to get to the mass-equivalence of visible light, we must, however, skip a step—

. . .

STEP 71

0.000 000 000 000 000 000 000 000 000 000 000 01 kilograms (10^{-35} kg)

0.000 000 000 000 01 attograms (10^{-14} ag)
5.6 electron volts ($10^{0.75}$ ev)

This mass-level would represent the mass-equivalence of photons of visible light, and is about one tenth that which may possibly represent the rest mass of a neutrino.

We could continue descending the scale through the mass-equivalence of the photons of infrared light and radio waves of longer and longer wavelengths, but that would add little to what we have already done, and we have gone far enough.

On the ladder of mass, we have progressed through 177 steps from the equivalent mass of a photon of visible light at the bottom, to the mass of the entire Universe on the top, all this covering 88 1/2 orders of magnitude.

The Ladder of Density

UPWARD

UPWARD

STEP 1

1 kilogram per litre (10^0 kg/L)

One can't help but notice that volume and mass do not always match, although we might suppose that if A is larger then B, then A should be both more voluminous and more massive than B. That, of course, is not always so.

To take an extreme case, Sirius B, the white-dwarf companion of Sirius, has a volume less than that of Earth, yet it has a mass that is several hundred thousand times as great as that of Earth. In fact, tiny Sirius B has a mass every bit as large as that of the much more voluminous Sun.

In other words, it is possible for a mass of material to be packed tightly into a small volume in one case, and loosely into a large volume in another. It is not enough, therefore, to speak of volume alone, or mass alone, if one wishes to know all there is to know about an object. We must also use a measure that describes how much mass is packed into a given volume of the object in question.

This is "mass per volume," and the quantity is called "density." Thus, Sirius B may be smaller than the Earth in volume, but Sirius B is enormously more dense than Earth is, and that accounts for the unexpectedly large mass of Sirius B.

In the SI version of the metric system, the basic unit of mass is the kilogram, and the basic unit of volume is the cubic metre. The logical unit for density would therefore be the number of kilograms of mass contained in a cubic metre of that substance. This could be spoken of as "kilograms per cubic metre" and symbolized as "kg/m^3."

Suppose, though, we consider the density of water. After all, we are often interested in whether an object is denser than water, or less dense than water.

(Laymen will very often say "heavier than water" or "lighter than water," but this is not what they really mean. Iron is "heavier" than water and wood is "lighter" than water, but an iron nail is lighter than a bathtub's worth of water, and a wooden beam is heavier than a pail's worth of water. What we really mean is that a particular substance is heavier or lighter than the *same volume* of water, and this is taken into account in the definition of density, so we should say "denser" and "less dense.")

An object that is denser than water will sink if placed in water, while an object that is less dense than water will float if placed in it. This is very important in a practical sense, so it is helpful if the units of density of water are so chosen as to give us some convenient figure.

In common American units, for instance, we would be apt to speak of density as so many "pounds per cubic foot." However, a cubic foot of water has a mass of 62.43 pounds, and that is an uncomfortably uneven number. For that matter, a cubic inch of water has a mass of 0.578 ounces, which is no better.

A cubic metre of water, on the other hand, weighs 1 000 kilograms, so that the density of water is 1 000 kilograms per cubic metre. The fact that it comes out to a round number is no accident. When the metric system was worked out, the units of mass and volume were deliberately chosen to give an even value for the density of water.

Still, 1 000 is a large number and perhaps we can do better.

The litre is a unit of volume that is permitted in the SI version of the metric system, as I mentioned earlier in the book. A litre is a cubic decimetre and, in volume, is equal to 1/1 000 of a cubic metre. Therefore, a litre of water must weigh 1/1 000 of 1 000 kilograms, or 1 kilogram. The density of water, then, is 1 kilogram per litre, and we can't ask for a more convenient figure than that.

When water freezes to ice, it turns out that the solid form is less dense than the liquid form. Ice has a density of 0.917 kilograms per litre and, therefore, floats on water. (Ice is exceptional in this respect. Most substances are somewhat denser in solid form than in liquid form.)

Water is a chemical combination of oxygen and hydrogen (H_2O). A nitrogen atom will combine with three hydrogen atoms to form "ammonia" (NH_3), and a carbon atom will combine with four hydrogen atoms to form "methane" (CH_4). The density of liquid ammonia is 0.817 kilograms per litre, while that of liquid methane is 0.415 kilograms per litre. In these two cases, the solid forms are somewhat denser than the liquid forms, but, even so, are less dense than liquid water.

Another common substance is carbon dioxide, a combination of a carbon atom with two oxygen atoms (CO_2). Liquid carbon dioxide has a density of 1.1 kilograms per litre, and solid carbon dioxide has one of 1.56 kilograms per litre. Both as liquid and as solid, carbon dioxide is denser than liquid water.

The solid forms of ammonia, methane, and carbon dioxide are each white and brittle, and have very much the appearance of ice. For that reason, these and other substances of the sort are referred to as "ices" or "icy materials." Ordinary ice is sometimes referred to as "water-ice" to distinguish it from the others.

In general, mixtures of ices have a density in the neighborhood of 1 kilogram per litre.

Many of the worlds of the Solar system consist chiefly of icy materials so that their overall density is in the neighborhood of 1 kilogram per litre. Thus, various satellites of Saturn have densities of from 1.0 to 1.4 kilograms per litre.

Titan, the largest satellite of Saturn, has a density of about 1.9 kilograms per litre. There are two reasons for this. In the first place, the more massive an object, the larger its gravitational field is likely to be, and the more that field will tend to pull the mass of the object inward and force it to take on a more compact arrangement, thus increasing its density. We would therefore expect the innermost regions of Titan to be compressed into greater density than the outer layers even if their basic composition were the same. That would increase the overall density. Secondly, of course, it may be that Titan has a considerable admixture of substances that are significantly denser than the ices.

Similarly, of Jupiter's satellites, Callisto has a density of 1.81 kilograms per litre, and Ganymede, one of 1.93 kilograms per litre.

The giant planets have a considerable content of hydrogen and helium, which are less dense than the icy materials. On the other hand, the gravitational fields of the giant planets are huge enough to produce an enormous compressing effect. The result is that Neptune has an overall density of 1.66 kilograms per litre; Jupiter, one of 1.314 kilograms per litre; and Uranus, one of 1.19 kilograms per litre.

The Sun, too, has an overall density in this range, the figure being 1.409 kilograms per litre. Here, there happens to be a balance between the tendency of the huge mass and gravitational field to compress the Sun and make it denser, and the tendency of the high temperature to expand it and make it less dense. The Sun happens to end up in the Step-1 density-region.

Coming closer to home, there are some metals in the Step-1 density-region. Magnesium has a density of 1.738 kilograms per litre, and calcium, one of 1.55 kilograms per litre. Of the common metals, sodium and potassium are actually less dense than water, with densities of 0.971 and 0.862 kilograms per litre, respectively.

STEP 2

3.16 kilograms per litre ($10^{0.5}$ kg/L)

Of the common elements, sulfur has a density of 2 kilograms per litre. Carbon varies in density according to the arrangement of its atoms. The most compact arrangement is found in diamond, which is therefore the densest form of carbon. It has a density of up to 3.5 kilograms per litre.

Of the metals, aluminum has a density of 2.7 kilograms per litre; barium, 3.5 kilograms per litre; and scandium, 3.0 kilograms per litre.

Most of the rocky materials that make up the Earth's crust have densities close to the Step-2 range. The average density of the Earth's crust is about 2.8 kilograms per litre. The Earth's mantle, which is deeper beneath the surface and has the weight of the upper layers pulling down upon it, thanks to Earth's gravitational field, is denser than the crust, reaching a figure of 4.5 kilograms per litre.

The average density of some sizable worlds is in this region, too. Thus, the average density of the Moon is 3.34 kilograms per litre, which shows that it is probably essentially rocky in nature through and through.

Two of the Galilean satellites are similar. The average density of Europa is 3.04 kilograms per litre, and that of Io is 3.55 kilograms per litre. Jupiter-probes have shown us that Europa is covered by a worldwide layer of ice, but this layer cannot be very thick, or Europa's density would be only half what it is.

Among the planets, the density of Mars is also low enough, 3.94 kilograms per litre, to indicate an essentially rocky structure.

STEP 3

10 kilograms per litre (10^1 kg/L)

On the whole, the denser metals are denser than any of the rocky materials, and at the Step-3 density-level, we have already left the rocks behind. Iron has a density of 7.874 kilograms per litre, and the similar metal, nickel, has one of 8.902 kilograms per litre.

It is thought that the interior of the Earth, its "core," is a nine to one mixture of iron and nickel, one that would have a density of about 8 kilograms per litre, if it were present in small amounts on Earth's surface. However, the core is compressed by the weight of the mantle above it. For that reason, the overall density of the Earth's core is about 10.7 kilograms per litre, with a mark of perhaps 11.5 kilograms per litre at the very center.

As a result of the presence of the metallic core, the overall density of Earth is 5.52 kilograms per litre, the highest overall density of any sizable object in the Solar system. The only other large objects approaching Earth's density are Mercury with an overall density of 5.42 kilograms per litre; and Venus, with one of 5.25 kilograms per litre. Both bodies must resemble Earth in structure, and each must possess a sizable metallic core. (Some relatively small meteors, largely nickel-iron in composition, have an overall density of about 8 kilograms per litre.)

Jupiter has a much larger mass than Earth, and a much more intense gravitational field. If it resembled Earth in structure, it would be bound to have a core denser still. Jupiter, however, is made up of material that is very low in density under ordinary conditions, and all of Jupiter's gravitational compression brings its density at its center to not more than 4.21 kilograms per litre.

There are metals that are closer to the Step-3 density-level than iron and nickel are. The density of thulium, lutetium, and bismuth are, respectively, 9.33, 9.75, and 9.87 kilograms per litre.

These are all on the low side of the range, but on the high side, we have molybdenum, silver, lead, and thallium, with densities, respectively, of 10.22, 10.50, 11.35, and 11.85 kilograms per litre.

STEP 4

31.6 kilograms per litre ($10^{1.5}$ kg/L)

To have a really dense substance, two requirements must be met. First, the atoms must be massive, and second they must be put together in a compact arrangement.

Because of the first requirement, the really dense substances on Earth are found among the elements with a high "atomic weight." Because of the second, one element may be slightly higher in atomic weight than another, and yet slightly lower in density. Thus, the highest-density metals on Earth are listed below, and, in parenthesis, the masses of the individual atoms of those metals are given in daltons:

Uranium (238)	18.95 kilograms per litre
Tungsten (183.9)	18.95 kilograms per litre
Gold (197)	19.32 kilograms per litre
Rhenium (186.2)	21.02 kilograms per litre
Platinum (195)	21.45 kilograms per litre
Iridium (192.2)	22.42 kilograms per litre
Osmium (190.2)	22.57 kilograms per litre

Osmium is the densest substance on the face of the Earth, and even it does not reach the Step-4 density-level.

Materials in liquids and solids, under conditions that prevail on Earth's surface, consist of intact atoms that are virtually in contact. These atoms consist of a tiny nucleus at the very center of the atom, a nucleus that contains almost all the mass of the atom, and electrons in the outer regions that take up most of the volume of the atom while contributing very little of the mass.

A great deal of compression can force the atoms closer together, increasing the density by decreasing the volume of the individual atoms. If pure osmium were at the core of the Earth, the overlying pressures of Earth's outer core, mantle, crust, hydrosphere, and atmosphere might increase its density to just about the Step-4 mark.

STEP 5

100 kilograms per litre (10^2 kg/L)

The Sun is much more massive than any planet, even Jupiter. Naturally, then, it might be expected to compress its materials at the center to much greater densities that those which exist at the center of Jupiter.

The Sun, like Jupiter, is mostly hydrogen, which is, under ordinary Earthly conditions, not at all dense. At the Sun's core, however, the gravitationally impelled downward pressure of all the layers above would put enough stress on the atoms of hydrogen to smash them: to break through the electron structure in the outer reaches of the atom. The atomic nuclei are then driven much closer together than is possible when the atoms are intact. This means that much more mass is forced into a given volume and the density goes up—way up.

As we shall see later in the book, such a smashing of atoms at the center of a massive object produces high temperatures which expand the object and which lower the density that would otherwise be produced. Even so, the density of the material at the center of the Sun is estimated to be about 160 kilograms per litre, about seven times the density of osmium at the Earth's surface.

STEP 6

316 kilograms per litre ($10^{2.5}$ kg/L)

The more massive a star is, the denser its core may be expected to be. Direct measurements of the density are impossible, but there are mathematical models that allow the density at the core to be calculated (though one can't be completely certain, in the absence of direct measurement, that a particular model is necessarily correct).

According to one model, we might expect a star like Vega, with three times the mass of the Sun, to have a central density of about 300 kilograms per litre, nearly twice that of the Sun, and nearly at the Step-6 density-level, too.

STEP 7

1 000 kilograms per litre (10^{3} kg/L)
1 tonne per litre (10^{0} t/L)

A star like Achernar, with six times the mass of the Sun, might have a central density of about 730 kilograms per litre, 4.5 times that of the Sun and within hailing distance of the Step-7 density-level.

STEP 8

3 160 kilograms per litre ($10^{3.5}$ kg/L)
3.16 tonnes per litre ($10^{0.5}$ t/L)

A star like Alpha Crucis (the brightest star in the constellation of the Southern Cross), with perhaps 16 times the mass of the Sun, might have a central density of about 2.6 megagrams (or tonnes) per litre, 18.6 times that of the Sun, and not too far below the Step-8 density-level.

STEP 9

10 000 kilograms per litre (10^4 kg/L)
10 tonnes per litre (10^1 t/L)

The most massive stars, with masses 30 times that of the Sun and more, might be expected to have central densities of from 7 to 10 tonnes per litre.

Such central densities exist in stars that are on the main sequence: that is, stars which derive their energy from the steady fusion of hydrogen at the core (as does our Sun). As such stars grow older, more and more of the hydrogen nuclei are fused into helium so that the core tends to grow denser. Eventually, when enough of the hydrogen nuclei are used up, conditions at the core allow the helium nuclei to be fused into still more complicated nuclei, and the core grows denser still.

One of the results of helium fusion is that the star is caused to swell into a "red giant." The cores of helium-fusing red giants are considerably denser than the cores of hydrogen-fusing main-sequence stars, and in order to represent it we must skip three steps.

. . .

STEP 12

316 000 kilograms per litre ($10^{5.5}$ kg/L)
316 tonnes per litre ($10^{2.5}$ t/L)

The more massive a star is to begin with, the larger the size of the red giant into which it will swell, and the greater the central densities it will develop. With Step 12, we are in the realm of the central densities of red giants, and a really large one could have a central density anywhere from 300 to 600 tonnes per litre.

Eventually, though, a red giant uses up so much of the fusible material in its core that it simply can no longer produce enough energy to keep itself expanded against its own gravitational pull, and it then collapses. The gravitational field intensifies as the star shrinks in size (especially if it does so without too much loss of mass), and the overall density rises rapidly as the atomic nuclei in the star come closer together. If the red giant is not very massive to begin with, it does collapse with very little mass loss, and becomes an object the size of Earth, or somewhat smaller. We then have a "white dwarf."

To represent the average density of a white dwarf, we must skip a step—

. . .

STEP 14

3 160 000 kilograms per litre ($10^{6.5}$ kg/L)
3.16 kilotonnes per litre ($10^{0.5}$ kt/L)

The most familiar white dwarf is Sirius B. It was the first one to be discovered and is the most studied. Its average density is about 2.9 kilotonnes per litre, which puts it almost at the Step-14 level. That average density is 530 000 times the average density of the Earth, and nearly 130 000 times the density of osmium on Earth's surface.

Of course, that is only its average density. Naturally, Sirius B is less dense than average in its outermost layers, but more dense than average in its deeper layers. Again, we will skip a step—

. . .

STEP 16

31 600 000 kilograms per litre ($10^{7.5}$ kg/L)
31.6 kilotonnes per litre ($10^{1.5}$ kt/L)

The central density of Sirius B is estimated to be about 33 kilotonnes per litre.

Even at such a density, the atomic nuclei are moving about freely. The density is over a million times as great as it would be in intact atoms, but the nuclei are so small that they are far from being in contact.

If density rose to the point where the nuclei *were* in contact, then the density of the object in which this took place would be equal to the density of atomic nuclei themselves.

If a star is sufficiently massive to begin with, then, at the time of collapse, the mounting intensity of the gravitational field would drive the nuclei into contact, at which point all the protons and electrons that had existed would combine to form neutrons, and only neutrons would be left. The result would be a "neutron star." To express the density of a neutron star, a large number of steps must be skipped—

. . .

STEP 33

10 000 000 000 000 000 kilograms per litre (10^{16} kg/L)
10 teratonnes per litre (10^{1} Tt/L)

This—ten quadrillion kilograms per litre—is the approximate density of neutron stars—and of atomic nuclei. Some 300 neutron stars have now been detected, and they are indeed exotic denizens of the astronomical zoo.

Yet it is possible for a really massive star to collapse so forcefully that even the resistance of neutrons in contact is not sufficient to withstand the rapidly intensifying gravitational compression. The neutrons themselves collapse, and the star continues to compress toward the zero point with nothing to stop it.

The result is a "black hole" in which, as size shrinks to zero, gravitational intensity in the neighborhood of the surface of the black hole increases without limit—and so does density.

In other words, we can envisage an endless series of steps upward from Step 33, all lying within the density-range of black holes. Since it is no use trying to list an endless series of density-steps, we will put an end to the upward ladder of density here.

The Ladder of Density

DOWNWARD

DOWNWARD

STEP 1

1 kilogram per litre (10^0 kg/L)

We are back to the density of water, as we make ready to descend the ladder of density.

Since living tissue is largely water, the overall density of human beings (and of many other forms of life) is at the Step-1 density-level. That is why we can manage the "dead man's float": resting horizontally on a water surface and drifting along it with just a small part of our body above it.

STEP 2

0.316 kilograms per litre ($10^{-0.5}$ kg/L)
3.16 hectograms per litre ($10^{0.5}$ hg/L)

Ethyl alcohol (the common "alcohol" of the beverages so many people drink) has a density of 7.9 hectograms per litre. It would, in principle, float on top of the water surface, if added to water—but it doesn't because it is completely soluble in water and, when added, simply mixes with it.

If, however, ethyl alcohol were placed in a sealed bag of thin aluminum foil, and the bag were dropped into water, the bag and its alcohol content would float.

Most organic compounds are lighter than water. Fats and oils have densities of about 9 hectograms per litre, and since they are insoluble in water, they will float upon it. The fatty content of a human being (and other animals) tends to lower the overall density below that of water, and to counterbalance those portions of the body —the bones, chiefly—that are denser than water. Stout people can, therefore, float more easily than thin people can.

Few liquids and solids, however, have densities at the Step-2 level. The least dense of all solids, under ordinary conditions, is the metal, lithium. (Its atoms are less massive than those of any other element that is solid at ordinary temperatures.) Its density is 5.34 hectograms per litre, so that it is only a little over half as dense as water, but it is still quite a bit above the Step-2 density-level.

To be sure, lithium can only be considered as the least dense solid, if we are talking of *continuous* solids, with all the atoms that make it up being in contact. It is also possible to consider solids that are actually loose networks, full of tiny air-filled holes and pores, sometimes so tiny as to be invisible to the unaided eye. The air adds very little to the mass of the solid so that the actually massive portion, the solid itself, is spread out over an uncharacteristically large volume and has an unusually low *overall* density.

Thus, if we try to think of a common solid that floats on water, we are sure to think of wood. Yet different kinds of wood vary widely in density, depending on how porous they are. Some types of wood, which are very compact and unporous, are actually denser than water and would sink if thrown into a pond. Ebony wood, for instance, can have a density of as much as 1.33 kilograms per litre. White pine, on the other hand, can have a density as low as 3.5 hectograms per litre, though the actual wood that makes it up is denser than water. Ordinarily, water does not penetrate the small air spaces in wood, but if, after long exposure, such penetration does take place, the wood becomes "waterlogged" and may sink.

STEP 3

0.1 kilograms per litre (10^{-1} *kg/L*)
1 hectogram per litre (10^{0} *hg/L*)

Solid materials at this density-stage are usually holes and pores for the most part. Cork (the bark of the cork tree) has a density of about 2.5 hectograms per litre, and there, the porousness of the substance is obvious. This is even truer of balsa wood, the density of which can be as low as 1.1 hectograms per litre. Balsa can be compressed with the fingers.

A Ping-Pong ball, which is a thin spherical shell of plastic enclosing air, has a radius of about 1.9 centimetres and a mass of about 2.5 grams. Just about all the mass is in the plastic shell, while almost all the volume consists of the enclosed air. The density is, therefore, a very low 0.9 hectograms per litre overall.

And yet we can find continuous solids at this density-level.

I said that lithium was the least dense of the continuous solids *under ordinary conditions.* There are, however, two elements with atoms less massive than those of lithium. These are helium and hydrogen, but they are gases under ordinary conditions. Under conditions of great cold, however, both helium and hydrogen can liquefy and even solidify. The liquids and solids formed by helium and hydrogen are the least dense continuous substancees that exist—or can exist. In both cases, the liquid is less dense than the solid.

Liquid helium has a density of 1.5 hectograms per litre, while liquid hydrogen has one of 0.7 hectograms per litre. A Ping-Pong ball made of solid hydrogen would have a smaller mass than an ordinary air-filled plastic Ping-Pong ball.

Gases are not continuous substances. Whereas liquids and solids are made up of atoms and molecules in contact, gases are made up of atoms and molecules that are not in contact and that are separated by space (much or little) containing no matter at all (vacuum).

Although the atoms of molecules of gases are as massive as those of many solids and liquids, the vacuum between the atoms and molecules of gases contributes no mass, so that though the volume increases, the mass does not. For that reason, gases, in general, have lower densities than solids and liquids do. And that is why solids that contain gas in pores, holes, or enclosed spaces are less dense than they would be if those did not exist and they extended continuously over the same volume.

Gases can be easily compressed; that is, the atoms and molecules making them up can be easily pushed more closely together by pressure. The density of gases therefore increases with pressure, and in extreme cases, the atoms and molecules can be pushed together until they are virtually in contact. In such a case, gases become as dense, more or less, as liquids and solids.

This can be done in the laboratory, but it can also take place in nature. Atmospheres of a planet are compressed under the pull of the planet's gravitational field. Those portions of the atmosphere near the planet's solid or liquid surface are compressed by the downward push of all the portions lying above. Where an atmosphere is very deep and a gravitational field is very intense, gas densities approaching those of liquids and solids are achieved. This undoubtedly happens in the atmosphere of the "gas giants," particularly Jupiter—and, of course, in stars, too.

Of the smaller planets, Venus has the most massive atmosphere. The gases near its surface are compressed to 90 times the density they would display near Earth's surface. Venus's atmosphere, which is mostly carbon dioxide, has a surface-density in the Step-3 range, with a value of about 1.7 hectograms per litre. That makes its density higher than that of liquid helium.

STEP 4

0.031 6 kilograms per litre ($10^{-1.5}$ kg/L)
3.16 dekagrams per litre ($10^{0.5}$ dag/L)

Earth's gravitational field has an intensity far less than that of a star, or of a giant planet, and its atmosphere is not particularly massive compared with that of Jupiter, or even with that of Venus. Therefore, the compression of the Earth's atmosphere at sea level is only moderate, and the density of air is quite low compared to even the least dense of the liquids and solids.

All gases in the neighborhood of Earth's surface experience the same compression. Therefore, the density of a particular gas, under sea level conditions, varies according to the mass of the atoms or molecules that make it up. The more massive the atoms or molecules, the denser the gas.

A gas with particularly massive molecules is uranium hexafluoride, each molecule of which is made up of a uranium atom and six fluorine atoms—UF_6. (Actually, it is not a gas under ordinary conditions, but boils more easily than water does, forming a vapor that is dense indeed for a gas.)

The uranium hexafluoride molecule has a mass of 352 daltons, compared to an average of 29 daltons for the molecules in air. The density of uranium hexafluoride vapor is therefore rather close to the Step-4 density-level, even though it is mostly vacuum. Under sea level conditions, it has a density of 1.6 dekagrams per litre, which is about a quarter the density of liquid hydrogen.

(On Venus, uranium hexafluoride vapor would be compressed to 90 times its density on Earth, so its Venusian density would be 1.4 kilograms per litre, and it would be denser than ebony wood.)

STEP 5

0.01 kilograms per litre (10^{-2} kg/L)
1 dekagram per litre (10^{0} dag/L)

At the Step-5 density-level, we find ourselves well within the range of the denser gases. Iodine vapor (with a molecule made up of two iodine atoms, I_2) has a density of 1.13 dekagrams per litre, about three fourths the density of uranium hexafluoride.

The gas radon is made up of single atoms, but very massive ones. It is the densest gas that is made up of single atoms; it is also the densest that is a gas at ordinary temperatures. (Uranium hexafluoride and iodine must be heated before they will produce vapors.) The density of radon is just about 1 dekagram per litre.

STEP 6

0.003 16 kilograms per litre ($10^{-2.5}$ kg/L)
3.16 grams per litre ($10^{0.5}$ g/L)

Sulfur tetrafluoride (each molecule containing one sulfur atom and four fluorine atoms, SF_4) has a density of about 4.82 grams per litre. Arsine (each molecule containing an arsenic atom and three hydrogen atoms, AsH_3) has a density of about 3.48 grams per litre.

STEP 7

0.001 kilograms per litre (10^{-3} kg/L)
1 gram per litre (10^{0} g/L)

We come now to the most familiar gases of all. Carbon dioxide, CO_2, the gas we (and all other forms of animal life) produce in respiration and the gas that green plants make use of in building their tissues, has a density of about 1.98 grams per litre.

Oxygen (with molecules made up of two oxygen atoms, O_2), which is the gas that supports animal life, has a density of 1.43 grams per litre. Nitrogen (with molecules made up of two nitrogen atoms, N_2) has a density of 1.25 grams per litre.

Air is a mixture of nitrogen and oxygen, in a proportion of nearly four to one in favor of the former. Air has a density, then, of about 1.29 grams per litre. The density of air, the most common gas, is 1/775 the density of water, the most common liquid. Air is about 1/55 the density of liquid hydrogen, the least dense of all continuous substances.

Large stars tend to be less dense than small stars, as the greater temperatures at the core of the larger stars tend to expand the volume mightily. The typical red giant star has an average density of 1 gram per litre, or just about the density of Earth's atmosphere at sea level. (Naturally, the core of such a star is enormously dense, and the outer regions are very little removed from a vacuum.)

STEP 8

0.000 316 kilograms per litre ($10^{-3.5}$ kg/L)
3.16 decigrams per litre ($10^{0.5}$ dg/L)

There are gases considerably less dense than air. For instance, the gas neon, made up of single neon atoms, has a density of 9 decigrams per litre, and is only three fourths the density of nitrogen, which is itself the less dense of the two chief components of air.

Water vapor, at ordinary temperatures, has a density of 8 decigrams per litre; ammonia has one of 7.6 decigrams per litre; and methane one of 7.1 decigrams per litre.

All these gases are substantially above the Step-8 level of density, however, and the truth is that there is no gas, under natural conditions on Earth's surface, that is close to the Step-8 density-level.

That does not mean, however, that we have come to an end by any means. Let us move on.

STEP 9

0.000 1 kilograms per litre (10^{-4} kg/L)
1 decigram per litre (10^{0} dg/L)

The two gases that are least dense are hydrogen and helium—which are made up of the smallest atoms of all.

Helium is made up of single helium atoms, and hydrogen is made up of molecules, each of which contains two hydrogen atoms (H_2). The density of helium is 1.78 decigrams per litre, while that of hydrogen is 0.9 decigrams per litre.

Helium is only one seventh as dense as air, so that it floats on air very readily (if not allowed to mix with it). It is used to fill balloons, from the toy ones sold at parades and other celebrations, to huge ones that carry human beings and instruments up into the stratosphere. Hydrogen, with only one fourteenth the density of air, is more effective still, but has the disadvantage of being inflammable, and even explosive, whereas helium will not burn or undergo any other kind of chemical reaction under any circumstances.

STEP 10

0.000 031 6 kilograms per litre ($10^{-4.5}$ kg/L)
3.16 centigrams per litre ($10^{0.5}$ cg/L)

Although hydrogen gas is the least dense substance that exists naturally on the Earth's surface, we can find lower densities if we leave Earth's surface.

If we move upward through the atmosphere, there is less and less of it that is higher still. There is, therefore, less and less weight of atmosphere above to compress the region below. Consequently, the air becomes less dense as we rise.

By the time we get some 25 kilometres above Earth's surface (three times as high as the peak of Mount Everest), the density of the air about us is in the neighborhood of 3 centigrams per litre, the Step-10 density-level. Long before we reach that height, of course, the air is insufficiently dense to support life, and human beings could only reach such heights alive if they were in sealed vehicles containing an atmosphere at normal density.

At this height, we are above the troposphere, that region of the lower atmosphere in which all the weather changes take place—in which there are clouds and precipitation. We are well into the stratosphere and are into that portion of it that is comparatively rich in ozone (a variety of oxygen, with three atoms to the molecule, O_3, that blocks the dangerous portion of the Sun's ultraviolet radiation). This region is sometimes called the "ozonosphere."

STEP 11

0.000 01 kilograms per litre (10^{-5} kg/L)
1 centigram per litre (10^{0} cg/L)

At a height of 35 kilometres above the Earth's surface, the density of the air is about 1 centigram per litre.

The planet Mars has a very thin atmosphere by Earthly standards, and the density of the Martian atmosphere (mostly carbon dioxide) at the planetary surface is only about 1 centigram per litre. The Martian atmosphere could not support Earthly animal life, in other words, even if it were pure oxygen, which it is not.

STEP 12

0.000 003 16 kilograms per litre ($10^{-5.5}$ kg/L)
3.16 milligrams per litre ($10^{0.5}$ mg/L)

The density of the Earth's atmosphere reaches the Step-12 level at a height of about 45 kilometres above the surface.

STEP 13

0.000 001 kilograms per litre (10^{-6} kg/L)
1 milligram per litre (10^{0} mg/L)

This is the density of Earth's atmosphere at a bit more than 50 kilometres above the Earth's surface. At this height, 98 percent of the atmosphere is below us. Above us would be the 2 percent of the "upper atmosphere."

The very largest red giant stars have an average density of 1 milligram per litre. Stars like Antares or Betelgeuse, which gleam brightly red in our sky, are actually large volumes of gaseous matter of less-than-stratospheric density. In fact, the density of the surface we see is far less than that of Earth's stratosphere. Of course, the core of such a red giant is, nevertheless, at white-dwarf densities.

We could continue moving down the scale of densities by reaching greater and greater heights above Earth's surface, but this would leave us, for the most part, with little interesting to say. Let us, therefore, skip steps where necessary—

. . .

STEP 15

0.000 000 1 kilograms per litre (10^{-7} kg/L)
100 micrograms per litre (10^{2} μg/L)

At a height of about 70 kilometres above Earth's surface, the density is about 100 micrograms per litre. At this height, many of the sparsely strewn atoms of the atmosphere have absorbed hard radiation from the Sun and have had electrons knocked out of them. The free electrons carry a negative electric charge, and the remaining atomic fragments, minus those ejected electrons, carry a positive electric charge. These atomic fragments are called "ions," and this region of the atmosphere is the "ionosphere."

The ionosphere reflects radio waves that are beyond the microwave region. This allows radio waves used in Earthly communication to reach far around the curve of Earth's surface as they bounce between the Earth's surface and the ionosphere. The ionosphere also reflects much of the radio wave radiation from the Universe generally and keeps it from reaching Earth's surface—

. . .

STEP 25

0.000 000 000 001 kilograms per litre (10^{-12} kg/L)
1 nanogram per litre (10^{0} ng/L)

At a height of 160 kilometres above the Earth's surface, the atmospheric density is 1 nanogram per litre. The less the density of any gas, the less the resistance to objects moving through it.

An object that is moving fast enough in a horizontal path above Earth's surface need not fall to that surface. Gravitational pull does cause it to drop, to be sure, but that drop (if the object is moving quickly enough in the horizontal direction) may just match the downward curve of the spherical surface. The moving object thus remains at a fixed distance above the curving surface, even though it is indeed falling. The object is "in orbit" about the Earth.

If, however, the object is so close to Earth's surface that it is passing through the atmosphere, each atom it strikes is speeded as a result of the collision. The atom gains momentum, and that momentum is lost by the speeding object. The amount of momentum lost to a single atom is totally insignificant, but atoms are hit by the trillions in every short interval of time, and the momentum of the object in orbit (and hence its speed) is slowly diminished. The object does not move fast enough to match the curve of the Earth, spirals inward, and eventually collides with the surface. Its orbit has "decayed."

The greater the height at which an orbit is established, the less dense the atmosphere, the fewer the atoms colliding with the moving object, and the longer it takes for the speed to diminish to the point of total orbital decay.

At a density of 1 nanogram per litre, air resistance is small enough to allow an orbiting satellite to remain in orbit for a considerable period of time, so that successful satellites usually approach no closer to Earth's surface than 160 kilometres.

. . .

STEP 31

0.000 000 000 000 001 kilograms per litre (10^{-15} kg/L)
1 picogram per litre (10^{0} pg/L)

At a height of about 450 kilometres above the Earth's surface, the density of the atmosphere has dropped to about 1 picogram per litre, a thousandth of the density at the 160-kilometre height.

Here we may expect orbiting satellites to remain in orbit for years —and yet not forever. Skylab, the orbiting space station on which three American astronauts remained for months at a time on three different occasions, was launched in 1973 in an orbit that varied from 400 to 425 kilometres above Earth's surface, and yet, by mid-1979, its orbit had decayed and it had crashed to Earth.

All atmospheres, even those much thicker than Earth's, become less and less dense with height, and eventually reach these trifling levels. Even the Solar atmosphere thins out.

At a height of about 2 megametres above the Sun's visible surface, we pass a transitional region above which is the Sun's upper atmosphere or "corona." At the lowermost portion of the corona, the density of the wisps of matter it contains is also in the 1 picogram per litre range—

. . .

STEP 37

0.000 000 000 000 000 001 kilograms per litre (10^{-18} kg/L)
1 femtogram per litre (10^0 fg/L)

At about 1 megametre above the Earth's surface, the density of the atmosphere has dropped another three orders of magnitude to 1 femtogram per litre.

At this height, we reach the portion of the atmosphere which is called the "exosphere," where the atoms and molecules of gas are so far apart, and so thinly spread, that the likelihood of their striking each other as they move about randomly is very small. They move virtually in independent orbits about the Earth.

While gases considerably denser than this are generally found in the neighborhood of planets and stars, there are thin wisps of gas also in certain regions of the space between the stars. These "interstellar clouds" are, at their densest, in the femtogram per litre stage, so that even the densest of such clouds represent nothing more than the equivalent of Earth's exosphere. Nevertheless, the interstellar clouds are so enormous in volume that their total mass is quite large, and they play important roles in cosmic evolution.

It is out of interstellar gas clouds that stars and their planetary systems condense, as our own Solar system did eons ago. How such condensation can take place out of such rarefied gas, with gravitational fields that are very attenuated, is something that astronomers are still trying to work out.

Then, too, it is now known that molecules exist in the gas clouds, including some surprisingly complex ones that are made up of seven or more atoms. How these could form when, at those densities, there are atomic collisions only very rarely, is another puzzle—

. . .

STEP 41

0.000 000 000 000 000 000 01 kilograms per litre (10^{-20} kg/L)
10 attograms per litre (10^{1} ag/L)

Gases of extremely low density do not exist only at appreciable distances above the surfaces of worlds, or in interstellar gas clouds. They can even exist on the very surface of the Earth, provided human beings form them. Over the last three centuries, scientists have learned to suck air out of enclosed volumes with greater and greater efficiency, thus forming vacuums.

The word "vacuum" is from the Latin word for "empty," and a human-made vacuum is certainly comparatively empty of matter—much more nearly empty than the surrounding environment. It is never *completely* empty, however. The best human-made vacuum has a density of about 10 attograms per litre, which is the naturally occurring density of Earth's atmosphere at a height of about 10 megametres above the surface.

At that height, we are into the lowest layer of the "magnetosphere," where there are streams of electrically charged particles moving along the lines of force of Earth's magnetic field.

At this step, it might be more convenient to begin to look at densities as representing the number of atoms per litre. The simplest atom is the hydrogen atom, and this is by far the most common atom in the Universe. If we suppose that very thin wisps of gas occurring in nature are entirely hydrogen, we will never be very far wrong, and in that case, we can say that a density of 10 attograms per litre would represent 6 000 000 hydrogen atoms per litre. We can symbolize "hydrogen atoms per litre" as "H/L"—

. . .

STEP 47

0.000 000 000 000 000 000 000 01 kilograms per litre (10^{-23} kg/L)
0.01 attograms per litre (10^{-2} ag/L)
6 000 hydrogen atoms per litre ($10^{3.78}$ H/L)

If we continue to move upward, the density of the atmosphere does not drop to zero. If we move away from the Earth to a distance equal to that of the Moon's orbit, we find that the density of the atmosphere tends to level off, because Earth itself, and the Moon, are actually moving around the Sun *within* faint wisps of the Sun's corona.

In other words, interplanetary space, at least in Earth's neighborhood, has a density of about 6 000 hydrogen atoms per litre. Objects moving in orbits closer to the Sun, as do Venus and Mercury, move through greater densities. Those moving in orbits farther from the Sun, as do Mars and the outer planets, move through lesser densities.

At such tiny gas densities (and considering the vast mass of even large asteroids), planetary orbits do not decay even over billions of years—

. . .

STEP 49

0.000 000 000 000 000 000 000 001 kilograms per litre (10^{-24} kg/L)
0.001 attograms per litre (10^{-3} ag/L)
600 hydrogen atoms per litre ($10^{2.78}$ H/L)

At this stage, we must leave the Solar system.

The Galaxy is filled with hundreds of billions of stars and, presumably, even vaster numbers of smaller, nonluminous objects. Nevertheless, these are spread apart so widely, over so enormous a volume, that the Galaxy is largely empty space.

If all the matter in the Galaxy were spread out evenly over all the vast volume of it, then it is estimated that the average density would prove to be only about 600 hydrogen atoms per litre. The Galaxy contains, on the average, only 1/10 000 as much matter in a given volume as the best vacuum that human beings can make—

. . .

STEP 51

0.000 000 000 000 000 000 000 000 1 kilograms per litre (10^{-25} *kg/L)*
0.000 1 attograms per litre (10^{-4} *ag/L)*
60 hydrogen atoms per litre ($10^{1.78}$ *H/L)*

Of course, matter in the Galaxy is *not* distributed evenly. Perhaps 90 percent of the matter is located within the stars and planets generally. Only 10 percent is distributed through the interstellar spaces.

The actual density of interstellar space, then, is, on the average, perhaps 60 hydrogen atoms per litre. This is equivalent to 60 000 hydrogen atoms per cubic metre (H/m^3), the unit of density to which we will now shift—

. . .

STEP 60

0.000 000 000 000 000 000 000 000 000 003 16 kilograms per litre ($10^{-29.5}$ *kg/L)*
0.000 000 003 16 attograms per litre ($10^{-8.5}$ *ag/L)*
2 hydrogen atoms per cubic metre ($10^{0.3}$ H/m^3*)*

Now we must leave the Galaxy and consider the entire Universe.

Galaxies are distributed so widely that there is more space outside galaxies than inside. If all the matter in all the galaxies in the Universe were smeared out evenly, then what would the average density be?

Actually, astronomers are not certain. One key figure is an average density of about 3 hydrogen atoms per cubic metre, a density just above the Step-60 level. If the density were any lower than that, the Universe would continue to expand forever. It would be an "open Universe." If it were any higher than that, it would eventually contract again and would be a "closed Universe."

All depends on whether the neutrinos have a tiny bit of mass or not. If they do, the Universe is probably closed; if not, the Universe is probably open.

Incidentally, 3 hydrogen atoms per cubic metre is equivalent to 3 000 000 000 hydrogen atoms per cubic kilometre (H/km^3) and it is to that unit that we will now shift—

. . .

STEP 65

0.000 000 000 000 000 000 000 000 000 000 01 kilograms per litre (10^{-32} kg/L)
0.000 000 000 01 attograms per litre (10^{-11} ag/L)
6 000 000 hydrogen atoms per cubic kilometre ($10^{6.78}$ H/km^3)

If the neutrinos are left out of account, astronomers suspect that the average density of the Universe may be in the neighborhood of 6 million hydrogen atoms per cubic kilometre, only 1/500 of the value required to close the Universe. This amounts to 1 hydrogen atom in 167 cubic metres or 167 000 litres. The Universe is simply an incredibly rarefied vacuum—on the average—

. . .

STEP 69

0.000 000 000 000 000 000 000 000 000 000 000 1 kilograms per litre (10^{-34} kg/L)
0.000 000 000 000 1 attograms per litre (10^{-13} ag/L)
60 000 hydrogen atoms per cubic kilometre ($10^{4.78}$ H/km^3)

Of course, about 99 percent of the matter of the Universe is likely to be contained in the various galaxies. In that case, the density of intergalactic space, on the average, would be in the neighborhood of 60 000 hydrogen atoms per cubic kilometre. This would amount to 1 hydrogen atom in about 17 000 cubic metres, or 17 000 000 litres, and lower than this density we cannot go.

Thus, in 100 steps covering 50 orders of magnitude, we have passed from the unimaginably low density of intergalactic space to the unimaginably high density of a neutron star. Even so, we have left out of account the density ranges of black holes, and that might make the number of steps infinite.

R. JONES

The Ladder of Pressure

UPWARD

UPWARD

STEP 1

1 000 pascals (10^3 Pa)
1 kilopascal (10^0 kPa)

Pressure is a force exerted over a particular area.

The most common force is weight. In speaking of pressure, then, we might speak of the pressure produced by the weight (at sea level on Earth) of a column of mercury of a given height resting on a given area. We might speak of the pressure of, let us say, a column of mercury 7.6 decimetres high resting on an area of 1 square metre. Such a column of mercury would exert a pressure of 10 300 kilograms (of weight) per square metre.

Although it is common to speak of pressure as weight per area, as so many pounds per square inch or as so many grams per square centimetre, that is not proper according to the SI version of the metric system. Grams and kilograms (or pounds, for that matter) are units of mass, and should not be used for weight. Weight is a force, and in the SI version, force is measured in "newtons" (named for Isaac Newton, who first worked out the correct mathematical concept of a force).

One newton is 1 kilogram-metre per second per second. One newton, in other words, is the amount of force required to impart to a mass of 1 kilogram an acceleration of 1 metre per second per second. Under the steady and continuing force of 1 newton, something with a mass of 1 kilogram, starting from rest, will (if we neglect friction and air resistance) be moving at a speed of 1 metre per second at the end of 1 second, 2 metres per second at the end of 2 seconds, 3 metres per second at the end of 3 seconds, and so on.

It is possible to calculate that a column of mercury 7.6 decimetres high will exert a pressure, due to its own weight, of 101 325 newtons per square metre (N/m^2), or 101.325 kilonewtons per square metre (kN/m^2).

In the SI version, 1 newton per square metre is, for brevity's sake, referred to as 1 pascal (pronounced pas-KAL, and symbolized "Pa") in honor of Blaise Pascal, a French physicist who made important discoveries in connection with air pressure. We can say, then, that a column of mercury 7.6 decimetres high exerts a pressure of 101.325 kilopascals (kPa).

A pascal is a very small unit of pressure, since it is that exerted by a column of mercury 7.5 micrometres high. Since water is 1/13.6 as dense as mercury, a column of water 13.6×7.5, or 102 micrometres high (and even that is a thin film not really visible to the unaided eye), will also produce 1 pascal of pressure. A kilopascal of pressure will be produced by a column of mercury 7.5 millimetres high or a column of water 1.02 decimetres high.

For convenience' sake, we will start the ladder at 1 kilopascal.

This puts us in the range of atmospheric pressure on the surface of the planet Mars. The Martian atmosphere is only about a hundredth as dense as Earth's atmosphere is. What's more, the Martian atmosphere is pulled downward by the Martian surface gravity, which is only two fifths as intense as Earth's surface gravity is. Consequently, the pressure of the Martian atmosphere at the surface of Mars is only about 0.8 kilopascals at the most.

When the Martian temperature is at its lowest (when Mars is farthest from the sun), enough carbon dioxide—which makes up 95 percent of the Martian atmosphere—freezes out to reduce the atmospheric pressure to 0.5 kilopascals.

The sea-level pressure of Earth's atmosphere is much higher than this, but as we move upward from the surface, more and more of the atmosphere is below us, and less and less above us, so that the pressure upon us sinks. At a height of 50 kilometres above Earth's sea level, atmospheric pressure is about 1 kilopascal.

STEP 2

3 160 pascals ($10^{3.5}$ Pa)
3.16 kilopascals ($10^{0.5}$ kPa)

As one moves downward through Earth's atmosphere, the pressure rises. At 25 kilometres above Earth's surface, the atmospheric pressure is 3.16 kilopascals.

STEP 3

10 000 pascals (10^4 Pa)
10 kilopascals (10^1 kPa)

At 15 kilometres above Earth's surface (a height 1.7 times that of the peak of Mount Everest), the atmospheric pressure is 10 kilopascals.

STEP 4

31 600 pascals ($10^{4.5}$ Pa)
31.6 kilopascals ($10^{1.5}$ kPa)

At 8 kilometres above Earth's surface (about the height of the peak of Mount Everest), the atmospheric pressure is close to the 31.6 kilopascal mark.

STEP 5

100 000 pascals (10^5 Pa)
100 kilopascals (10^2 kPa)
1 atmosphere (10^0 atm)

At the Step-5 pressure-level, we are just about at the value of Earth's standard atmospheric pressure at sea level. This is, actually, 101.325 kilopascals. It is why I started this section by dealing with a column of mercury 7.6 decimetres high. It is that which produces a "standard atmosphere" of pressure. The SI version permits the use of "1 standard atmosphere" as a unit of pressure equal to 101.325 kilopascals, and it is symbolized "atm." I shall omit the word "standard" and, for convenience' sake in these ladders, set "1 atmosphere" equal to 100 kilopascals.

If we consider those worlds of the Solar system that are less massive than Earth and have less intense gravitational fields, two have atmospheric pressures greater than that of Earth.

One of these (and the lesser) is Saturn's large satellite, Titan, which has only one fiftieth the mass of Earth and a surface gravity only one ninth that of Earth's. Even so, thanks to its low temperature, Titan can retain a dense atmosphere (mostly nitrogen) that has a surface pressure of at least 160 kilopascals, or 1.6 atmospheres.

STEP 6

316 000 pascals ($10^{5.5}$ Pa)
316 kilopascals ($10^{2.5}$ kPa)
3.16 atmospheres ($10^{0.5}$ atm)

We have now passed well beyond the range of any pressure that Earth's atmosphere can exert.

Water can, however, also exert a pressure. If an object is immersed in water, the weight of the water above it will exert a pressure on all parts of its surface. The deeper it is in a lake or in the ocean, the greater the weight of water above it, and the greater the pressure exerted upon it.

A column of water 10.332 metres high would possess as much weight as a column of air of identical cross-sectional area, extending upward the many miles to the top of the atmosphere. Therefore, the pressure of the water in which an object is immersed is equal to 100 kilopascals, or 1 atmosphere, for every 10.332 metres of immersion.

If an object is 22.3 metres below the surface of a lake, it will experience upon its surface a pressure of 316 kilopascals (3.16 atmospheres), of which 216 kilopascals (2.16 atmospheres) will be the result of the weight of water above, and the remaining 100 kilopascals (1 atmosphere), the result of the air pressure transmitted through the water.

Since ocean water (containing dissolved salt) is 3 percent denser than the fresh water of lakes, an object need merely be immersed under 21.7 metres of sea-water to experience a pressure of 316 kilopascals (3.16 atmospheres).

STEP 7

1 000 000 pascals (10^{6} Pa)
1 megapascal (10^{0} MPa)
10 atmospheres (10^{1} atm)

An object immersed at a depth of 9 dekametres below the ocean surface is subjected to a pressure of 1 megapascal, or 10 atmospheres.

STEP 8

3 160 000 pascals ($10^{6.5}$ Pa)
3.16 megapascals ($10^{0.5}$ MPa)
31.6 atmospheres ($10^{1.5}$ atm)

An object immersed at a depth of 3 hectometres below the ocean surface is subjected to a pressure of about 3.16 megapascals (31.6 atmospheres).

STEP 9

10 000 000 pascals (10^7 Pa)
10 megapascals (10^1 MPa)
100 atmospheres (10^2 atm)

I said earlier that there were two objects smaller than Earth with an atmospheric pressure greater than ours. One is Titan, and the other is, of course, Venus. The pressure at Venus's surface, under a thick atmosphere that is mostly carbon dioxide, is about 9 megapascals (90 atmospheres).

Such a pressure can be found on Earth at a depth of about 8.9 hectometres below the ocean surface.

An actual pressure of 10 megapascals (100 atmospheres), which is the Step-9 pressure-level, exists at a depth of very nearly 1 kilometre under the ocean surface.

STEP 10

31 600 000 pascals ($10^{7.5}$ Pa)
31.6 megapascals ($10^{1.5}$ MPa)
316 atmospheres ($10^{2.5}$ atm)

A pressure of 31.6 megapascals (316 atmospheres) exists at a depth of about 3 kilometres below the ocean surface. This is approaching the average depth of the ocean, and we are now heading for the abyss.

STEP 11

100 000 000 pascals (10^8 Pa)
100 megapascals (10^2 MPa)
1 000 atmospheres (10^3 atm)

We are now at the pressure found in the deepest trenches of the ocean. The pressure at the very bottom of the Marianas trench in the Pacific Ocean is about 108.4 megapascals, or 1 084 atmospheres.

STEP 12

316 000 000 pascals ($10^{8.5}$ Pa)
316 megapascals ($10^{2.5}$ MPa)
3 160 atmospheres ($10^{3.5}$ atm)

We have now passed beyond air and sea on Earth, but there remains the solid body of the planet itself. The rocky substance making up the Earth's crust is denser than the watery ocean so that a given column of it weighs more and exerts a greater pressure than the ocean itself.

Rock located about 10 kilometres below the surface of the Earth's solid crust would, for instance, be under a pressure of about 316 megapascals, or 3 160 atmospheres. It would be under more than three times the pressure at the deepest part of the ocean floor, though that is about 11 kilometres below the ocean surface.

STEP 13

1 000 000 000 pascals (10^{9} Pa)
1 gigapascal (10^{0} GPa)
10 000 atmospheres (10^{4} atm)

Here we find ourselves about 36 kilometres below the solid surface of the Earth, reaching a pressure of 1 gigapascal, or 10 000 atmospheres. We are just about at the bottom of the Earth's crust and at the top of the underlying mantle.

STEP 14

3 160 000 000 pascals ($10^{9.5}$ Pa)
3.16 gigapascals ($10^{0.5}$ GPa)
31 600 atmospheres ($10^{4.5}$ atm)

We are now well within the Earth's mantle. At a depth of 100 kilometres below the solid surface of the Earth, the pressure is about 3.16 gigapascals, or 31 600 atmospheres.

STEP 15

10 000 000 000 pascals (10^{10} Pa)
10 gigapascals (10^{1} GPa)
100 000 atmospheres (10^{5} atm)

At a depth of 300 kilometres below the surface of the Earth, the pressure is about 10 gigapascals, or 100 000 atmospheres.

STEP 16

31 600 000 000 pascals ($10^{10.5}$ Pa)
31.6 gigapascals ($10^{1.5}$ GPa)
316 000 atmospheres ($10^{5.5}$ atm)

At a depth of 800 kilometres below the surface of the Earth, the pressure is about 31.6 gigapascals, or 316 000 atmospheres.

STEP 17

100 000 000 000 pascals (10^{11} Pa)
100 gigapascals (10^{2} GPa)
1 000 000 atmospheres (10^{6} atm)

At a depth of 2.2 megametres below the surface of the Earth, over a third of the way to Earth's center, the pressure is about 100 gigapascals, or 1 000 000 atmospheres.

STEP 18

316 000 000 000 pascals ($10^{11.5}$ Pa)
316 gigapascals ($10^{2.5}$ GPa)
3 160 000 atmospheres ($10^{6.5}$ atm)

We are beyond the mantle now and deep within the Earth's metallic core. At a depth of 5 gigametres below Earth's surface, the pressure is about 316 gigapascals, or 3 160 000 atmospheres.

Indeed, we are almost at the limit of what Earth can offer us in the way of pressure. At the Earth's very center, a depth of 6.371 gigametres, the pressure is about 370 gigapascals, or 3 700 000 atmospheres.

Of course, the four giant planets are each much more massive than Earth, and, in their interiors, much greater pressures must exist than can exist anywhere within the Earth.

In the case of Saturn, for instance, a pressure of 370 gigapascals, or 3 700 000 atmospheres, is reached at a depth of 30 gigametres below the visible surface of the planet. This is nearly five times as deep as suffices to produce the pressure on Earth, which seems strange in view of the fact that Saturn's gravitational field is nearly a hundred times as intense as Earth's. However, the composition of

Saturn's outer layers, which are almost entirely hydrogen, is much less dense than the rock and metal that composes Earth.

Then, too, whereas one can only probe a little over 6 gigametres into the Earth before reaching the center and the limit of depth, one can probe 30 gigametres below Saturn's surface and find that one is only halfway to that voluminous planet's center.

Jupiter, which is denser and even more voluminous than Saturn is, possesses internal pressures that rise more rapidly. A pressure of about 370 gigapascals is reached perhaps 20 gigametres below Jupiter's visible surface, and that is only a little over a quarter of the way to its center.

STEP 19

1 000 000 000 000 pascals (10^{12} Pa)
1 terapascal (10^{0} TPa)
10 000 000 atmospheres (10^{7} atm)

At the terapascal range, Uranus and Neptune are left far behind. Their central pressures are perhaps 0.7 terapascals, or 7 000 000 atmospheres (even so, twice that at Earth's center).

Perhaps 48 gigametres below Saturn's visible surface, the pressure is at 1 terapascal, or 10 000 000 atmospheres. By then, we are four fifths of the way to Saturn's center, and are inside what may be a rocky core.

STEP 20

3 160 000 000 000 pascals ($10^{12.5}$ Pa)
3.16 terapascals ($10^{0.5}$ TPa)
31 600 000 atmospheres ($10^{7.5}$ atm)

Some 60 gigametres below Jupiter's visible surface, we approach what may be the planet's rocky core, and reach a pressure of 3.16 terapascals, or 31 600 000 atmospheres. At Saturn's very center, the pressure may be as much as 5 terapascals, or 50 000 000 atmospheres.

STEP 21

10 000 000 000 000 pascals (10^{13} Pa)
10 terapascals (10^{1} TPa)
100 000 000 atmospheres (10^{8} atm)

By now, we leave even Jupiter behind, for at Jupiter's core, the pressure is perhaps not more than 8 terapascals, or 80 000 000 atmospheres, a little over 20 times that at Earth's center.

There is still, of course, the Sun. About 270 gigametres below the Sun's visible surface, the pressures exceed Jupiter's utmost, reaching the 10 terapascal or 100 000 000 atmosphere range.

STEP 22

31 600 000 000 000 pascals ($10^{13.5}$ Pa)
31.6 terapascals ($10^{1.5}$ TPa)
316 000 000 atmospheres ($10^{8.5}$ atm)

Some 340 gigametres below the Sun's visible surface, just about halfway to the center, the pressure reaches 31.6 terapascals, or 316 000 000 atmospheres (four times the pressure at Jupiter's center).

STEP 23

100 000 000 000 000 pascals (10^{14} Pa)
100 terapascals (10^{2} TPa)
1 000 000 000 atmospheres (10^{9} atm)

About 380 gigametres below the Sun's visible surface, the pressure reaches the 100 terapascal, or 1 000 000 000 atmosphere stage. This is 12 times the pressure at the center of Jupiter, and at about this pressure, the atoms themselves give way. Below this level, the Sun is no longer made up of intact atoms, but of atom fragments that can pack together much more tightly.

STEP 24

316 000 000 000 000 pascals ($10^{14.5}$ Pa)
316 terapascals ($10^{2.5}$ TPa)
3 160 000 000 atmospheres ($10^{9.5}$ atm)

This range of pressure is reached about 450 gigametres below the Sun's visible surface, or two thirds of the way to its center.

STEP 25

1 000 000 000 000 000 pascals (10^{15} Pa)
1 petapascal (10^{0} PPa)
10 000 000 000 atmospheres (10^{10} atm)

Such a pressure is reached about 500 gigametres below the Sun's visible surface, about seven tenths of the way to its center.

STEP 26

3 160 000 000 000 000 pascals ($10^{15.5}$ Pa)
3.16 petapascals ($10^{0.5}$ PPa)
31 600 000 000 atmospheres ($10^{10.5}$ atm)

At 540 gigametres below the Sun's visible surface, more than three fourths of the way to its center, we reach this range of pressure.

STEP 27

10 000 000 000 000 000 pascals (10^{16} Pa)
10 petapascals (10^{1} PPa)
100 000 000 000 atmospheres (10^{11} atm)

The 10 petapascal, or 100 000 000 000 atmosphere, mark is reached about 600 gigametres below the Sun's visible surface.

STEP 28

31 600 000 000 000 000 pascals ($10^{16.5}$ Pa)
31.6 petapascals ($10^{1.5}$ PPa)
316 000 000 000 atmospheres ($10^{11.5}$ atm)

At the very center of the Sun, the pressure is about 32 petapascals, or 320 000 000 000 atmospheres. This is 4 000 times that at Jupiter's center, and 80 000 times that at Earth's, and here we will stop. Stretching ahead of us are an infinite number of steps of increasing pressure at the centers of stars more massive than the Sun, of white dwarf stars, neutron stars, and black holes—the pressures increasing without limit in the last category. We will get nothing, from this point on, that we had not obtained in ascending the ladder of density.

What is more, if we return to Step 1 and attempt to descend the ladder of pressure, we will gain almost nothing that we had not obtained in the downward descent in density.

We will be satisfied, therefore, in having gone from the surface pressure of the Martian atmosphere to the central pressure of the Sun in 27 steps, covering 13 1/2 orders of magnitude.

We will next go on to something entirely new, involving in no way either length or mass.

R. Jones

The Ladder of Time

UPWARD

UPWARD

STEP 1

1 second (10^0 s)

So far, we have dealt with two fundamental measurements: length and mass. We have also dealt with area, volume, and density, but area is the square of length, volume is the cube of length, and density is mass divided by volume. All these measures therefore boil down to length, mass, or a combination of the two.

What about pressure, though?—It is a force exerted against an area, and an area is the square of a length.

Yes, but what is a force?—Force is a mass multiplied by an acceleration, and mass is one of the fundamentals we have dealt with.

Yes, but what is acceleration?—Acceleration is a change of speed with time, and speed is a change of distance traveled (a length) with time.

In the definition of acceleration, time is mentioned twice, and you will remember that at the beginning of the section on pressure I gave the units of 1 newton (the basic unit of force) as 1 kilogram-metre per second per second. That "per second per second" is *again* a double mention of time for "second" is a unit of time.

In dealing with pressure, then, we have been dealing with time, although that certainly wasn't obvious. Let us now, therefore, go on to deal with time directly. It is certainly time we did that.

In dealing with time, we encounter a new phenomenon—a type of measurement that has defeated the metric system.

The common units of length which, in every nation, were invented in the uncertain past, were replaced by the metre together with its multiples and divisions. Similarly, the time-honored units of mass were replaced by the kilogram with its multiples and divisions.

Where time is concerned, however, the fundamental units were invented by the Sumerians (who dwelt in what is now Iraq) some 40 centuries ago or more. The Sumerian units spread outward and were eventually used throughout Europe, and are today used throughout the world.

The Sumerian units have *not* been replaced by the metric system and were never even challenged by it. The Sumerian unit "second" is accepted as the fundamental unit of time in the SI version of the metric system. It is symbolized as "s."

The second is a small unit of time and isn't used much in ordinary life. (Human beings are very poor in estimating periods of time, in any case.) So short is a second that human beings often think of it as a mere instant.

Actually, a good method of counting seconds is to say, "a thousand and one," "a thousand and two," "a thousand and three," and so on in a brisk but unhurried way. Don't race and don't drawl and you'll be pretty close to the mark.

In short, one second is the time it takes to articulate five syllables when most people speak in their usual fashion. (I can say "Isaac Asimov" in one second when I speak normally and don't pause between the first and last name.) This means that if you count, the best thing to do is to count up to "a thousand and six" and start over. "A thousand and seven" has an extra syllable and you must slur it, and after "a thousand and twelve" there are nothing but extra syllables.

Another way of counting the seconds without a mechanical device is to use your pulse. I have just taken my pulse, and I find that there were sixty-three beats in sixty seconds, which means that the time interval between two beats was just about 0.95 seconds—quite good.

Pulsebeats are erratic, however. The interval between beats will shorten if you are active, emotional, or just nervous. My own experience is that merely concentrating on my pulsebeat produces enough tension, no matter how calm I try to remain, to speed it up a bit.

Then, too, the pulsebeat is more rapid the smaller the human being, and the beat-per-second is more apt to be correct for adult males of normal size than for others. The adult female, being smaller than the adult male, on the average, has a more rapid pulsebeat, and the time interval between two of her pulsebeats is likely to be as little as 0.75 seconds. Children's pulsebeats are more rapid still, and at birth, the interval between pulsebeats is only 0.45 seconds.

Anything human or, more generally, living is a bit too erratic to be relied on as a measure of time interval, actually. We must find something better than speech or pulsebeat.

There is, for instance, the beat of a pendulum. A pendulum of a given length swings back and forth in a fixed time regardless of how long or how short the arc of the swing is. At least, the time is *almost* fixed. It varies slightly with the strength of Earth's gravitational attraction upon it and, therefore, with height above sea level. It also varies with the temperature. It also varies slightly with the size of the swing if the pendulum marks out the arc of a circle, as it naturally does, instead of the arc of a curve called the cycloid, as it can be made to do.

The pendulum came into use only in the seventeenth century, however. Before that, use had to be made of any continuous change that seemed to be proceeding at a constant rate, such as the dripping of water, or the sifting of sand, through a tiny orifice; or the burning of a candle; or the creeping of a shadow along the ground as the day progresses.

None of these methods is very accurate, but throughout history, they have been used in default of anything else. In fact, there are still ornamental sundials in some gardens, which work as well as they ever did; and we still use little devices in which sand drifts from an upper chamber to a lower one in a fixed time and allow that to guide the correct boiling of an egg, for instance.

It would be better if we could make use of some astronomical cycle, because these tend to be regular. The Earth turns on its axis in a fixed time; the Moon revolves about the Earth in a fixed time; the Earth revolves about the Sun; and so on. Each of these cycles defines a period of time, though all are much longer than the second. Still, the second can be defined as a fixed fraction of one of these much longer astronomical periods, and I will mention these definitions at the appropriate time.

As it happens, there is an astronomical cycle much shorter and potentially more convenient than any of those involving the Earth, Moon, and Sun. That is the rotation period of a neutron star.

The first neutron star discovered makes a complete turn in 1.337 301 09 seconds. These turns were marked out by pulses of radio waves that arrive with extraordinary regularity so that the objects were known as "pulsars" (an abbreviation of "pulsating stars") before they were discovered to be neutron stars.

A second could be defined as the time it takes that particular pulsar to make 0.747 774 71 of a turn. At least this could have been done after 1969, when that first neutron star was discovered.

All astronomical periods show certain irregularities, however (even neutron stars, whose cycles slowly grow longer with time). Scientists have therefore turned to the much, much faster cycles involving atoms which, as far as we know, do not change at all with time.

Atoms, in some of their activities, emit radiation of a certain frequency (so many waves or oscillations per second). Each different kind of atom will produce radiation characteristic of itself, but of no other kind of atom, and this frequency can be determined to a very high degree of precision. It is then necessary to agree on a particular type of atom producing radiation under set conditions and to define a second as being so many cycles of that radiation.

In 1964, the second was defined as 9 191 631 770 cycles of the radiation produced under certain specified conditions by an atom of cesium, of the isotopic variety known as cesium-133. In 1967, this was adopted as the definition of the second in the SI version of the metric system. This may seem cumbersome, but it produces a more precise and invariant standard for the exact length of duration of the second than does any other system scientists have ever worked out.

One more thing! In one second, a beam of light, traveling through a vacuum, will travel 299.792 456 2 megametres. This distance is equal to circumnavigating the Earth at the equator 7 1/2 times, or traveling just a bit over three fourths of the way to the Moon. In one second!

STEP 2

3.16 seconds ($10^{0.5}$ s)

Pulsars have different periods of rotation. Any particular pulsar, as it rotates, gradually loses energy by sending out streams of charged particles and radiation, and loses angular momentum in this way also. As the result of these losses, the period of rotation slowly lengthens and the intensity of radiation weakens. With time, therefore, pulsars grow steadily more difficult to detect.

The pulsar with the longest period known, so far, rotates in 3.755 seconds, just a bit above the Step-2 time-level.

It is this slow increase in period (by a value of a billionth of a second or so each day) that makes pulsars of limited usefulness as a standard for the definition of a second.

In 3.16 seconds, a beam of light can travel 947 megametres, or from the Earth to the Moon, and back—and then halfway to the Moon again. Or it can travel, in this time, from Jupiter to its largest satellite, Ganymede.

Let us bring matters down to everyday affairs. In music, the usual "quarter note" is held for a second. A "whole note" is therefore held for four seconds. In singing hymns, the final "amen" is usually shown as two whole notes. In singing "amen," then, the two syllables are held for four seconds each.

STEP 3

10 seconds (10^1 s)

Ten seconds is *not* referred to as a "dekasecond." The usual prefixes of the metric system are not commonly used in units of time greater than a second. There are other ways of handling that, as we shall see.

In 10 seconds, a professional sprinter can run up to a hundred metres from a kneeling start. The world record for the hundred-metre dash is, in fact, 9.95 seconds, and it was set in 1968.

In 10 seconds, on the other hand, a beam of light can travel nearly three fourths of the way around the Sun's equator.

STEP 4

31.6 seconds ($10^{1.5}$ s)

A duration of 31.6 seconds is a little over half a "minute" (symbolized "min"), since 60 seconds equals 1 minute. This is so by a convention first established by the Sumerians.

Why 60? The Sumerians had not yet developed good mathematical techniques for handling fractions, and it was particularly convenient for them to make use of numbers that could be divided evenly in a number of ways, so that the necessity of using fractions would be minimized. The number 10, which is a natural one for building higher units because that is the number of fingers on one's two hands, can be divided evenly by only 2 and 5. The number 60, on the other hand, can be divided evenly by 2, 3, 4, 5, and 6. It is the smallest number that can be divided evenly by every digit smaller than 7. (It can also be divided evenly by 10, 12, 15, 20, and 30.)

Nowadays, of course, fractions hold no terrors to those involved in arithmetical computations (well, to *some* of those so involved), and there have been suggestions that units of time be decimalized; that minutes be divided into 100 "metric seconds," for instance. However, it is unlikely that such a change can ever be accomplished.

As an example of a Step-4 time interval in daily life, most "commercials" that interrupt regularly scheduled television programs in prime-time are 30 seconds long. This makes it quite clear that 30 seconds is a longer interval of time than most of us are likely to assume.

STEP 5

100 seconds (10^2 s)
1.67 minutes ($10^{0.22}$ min)

In 100 seconds, the average educated adult can read about 500 words, and his heart will beat about 120 times, whether he is educated or not. In comparison, the heart of a shrew, the smallest mammal, will, in this interval, beat, under normal conditions, about 1 650 times.

STEP 6

316 seconds ($10^{2.5}$ s)
5.27 minutes ($100^{0.72}$ min)

This is about the time it takes for a human being to die of asphyxiation. In crime and horror movies, one person is sometimes shown using his two hands to choke a victim to death. This is almost never realistic. In the first place, it isn't that easy to choke a person, since the windpipe is stiffened with cartilage and a strong grip is therefore required to constrict it sufficiently. Second, the grip must be maintained for five minutes or so, if death is to be insured, and no movie is going to concentrate on the process for that long. Death is usually presented as taking place in a minute or less. (Incidentally, the reason for death is not that the heart or muscles or kidneys are deprived of oxygen. It is the oxygen-guzzling brain that is the weak point. After five minutes or so, the brain, deprived of oxygen, is permanently nonfunctional, and without the brain the rest of the body is useless.)

The Step-6 time interval allows for a reasonably long total eclipse of the Sun. On the average, the Moon is a trifle smaller in appearance than the Sun is, so that, more often than not, the Moon will not cover the entire Sun. When, however, the Moon is in the part of its orbit where it is closer than usual to the Earth, and therefore appears a bit larger than usual, while the Earth is in the part of its orbit where it is farther than usual from the Sun, so that the latter body appears smaller than usual, the Moon can cover all the Sun with something to spare. When the respective sizes are at their extreme, it can take as long as 7.5 minutes for the Sun, having slipped behind the Moon in the sky, to move far enough to begin to emerge at the other side.

STEP 7

1 000 seconds (10^3 s)
16.7 minutes ($10^{1.22}$ min)

The Earth's population has been increasing ever since *Homo sapiens* appeared on this planet. What's more, the rate at which the population increases itself tended to increase during the course of history. At the present time, the Earth's population is increasing by about 2 300 in a period equal to 16.7 minutes.

STEP 8

3 160 seconds ($10^{3.5}$ s)
52.7 minutes ($10^{1.72}$ min)

The Step-8 time interval is nearly an hour long, since 1 hour (symbolized as "hr") is equal to 60 minutes (or to $60 \times 60 = 3\ 600$ seconds). The hour, like the minute, can be used in the SI version of the metric system.

In ancient times, the hour was the smallest unit of time used in ordinary life, and the word comes from the Latin word for "time." It wasn't until the mechanical clock was developed in the Middle Ages, and, particularly, the pendulum clock in the 1650s, that it became possible to divide the hour with reasonable accuracy and make practical use of the sexagesimal division that Sumerians had worked out in theory.

The hour was divided into 60 equal parts, as I have already said, and each of those parts was spoken of as a "pars minuta prima" in Latin, meaning "first small part." This was shortened to "minute," and accented on the first syllable. (When accented on the second—my-NYOOT—it means "very small.")

When the minute was divided into 60 equal parts, each of these was called "pars minuta secunda" or "second small part," and that was shortened to "second."

The Sumerians divided the circumference of a circle into 360 equal units (where 360 is 60×6), each of which we now call a "degree." The degree, like the hour, is divided into 60 smaller units, and each of these into 60 still smaller units—and these are also referred to as minutes and seconds. To distinguish the divisions of a circle from the divisions of an hour, the former are referred to (when there is a chance of confusion) as "minutes of arc" and "seconds of arc."

Let's consider the Step-8 time interval in connection with the neutron. The neutron is one of the two particles that makes up the nucleus of an atom; the other is the proton. These together make up virtually all the mass of the Universe, as far as we know today (provided the neutrino does not turn out to have a small amount of mass).

The proton is stable. That is, if it is not interfered with by other particles, it will remain a proton forever (or, at least, for an extremely long time). The neutron is stable as long as it is present in the nucleus, but if it exists in isolation, it breaks down into a proton plus two other particles (the electron and the antineutrino) that have very little mass in comparison to the neutron and proton.

It is impossible to tell when a given neutron may break down. It may do so within a second, or not for a million years. However, if a large number of neutrons are present, then it is possible to predict, with considerable precision, how long a time must elapse before half will have broken down. We don't know exactly which neutrons will be included in that half, but half will then have broken down. The period of time during which this takes place is the "half-life," and the half-life of the neutron is just about 13 minutes.

Once half the neutrons have broken down, half of the half that remains will break down in the next 13 minutes, then half of the half that still remains will break down in the next 13 minutes, and so on.

Since 52.7 minutes is just about four times as long as 13 minutes, this means that only half of half of half of half, or one sixteenth, of the original number of neutrons will remain as neutrons after a Step-8 time interval. In 52.7 minutes, in other words, about 94 percent of any given large number of neutrons will have broken down.

STEP 9

10 000 seconds (10^4 s)
2.78 hours ($10^{0.44}$ hr)

At the Step-9 time-level, we can begin to be astronomical. Undoubtedly, there are some small asteroids that rotate about their axes in this time or less. Similarly, small objects in the immediate neighborhood of a planet could revolve about it in this time or less.

Thus, Yuri Gagarin, on April 12, 1961, was carried around the Earth on a rocketship in ballistic flight (moving naturally under the influence of gravity, just as a natural satellite would have done). He was at an average height of about 250 kilometres above Earth's surface, and his time of revolution was 1.8 hours.

STEP 10

31 600 seconds ($10^{4.5}$ s)
8.78 hours ($10^{0.94}$ hr)

This brings us to named astronomical objects. Phobos, the inner satellite of Mars, revolves about its planet in 7.65 hours. Amalthea, closer to Jupiter than the large Galileans, revolves about that planet in 11.74 hours. These two satellites, in their periods of revolution, bracket the Step-10 time-level.

Jupiter, the largest planet, has, as it happens, the shortest period of sidereal rotation (that is, rotation relative to the stars). It rotates (at the equator) in only 9.841 hours. Saturn, the second largest planet, rotates in 10.233 hours.

We have also reached the lifetimes of organisms large enough to be seen with the unaided eye. There are insects known as mayflies, which, in their adult stage, live only long enough to mate and lay eggs. They do not eat. In fact, their mouthparts are vestigial so that they cannot eat. Some species only live long enough to allow Phobos or Amalthea to revolve about their planets once, or to allow Jupiter or Saturn to rotate once about their axes.

Mayflies belong to the order of "Ephemerida," from Greek words meaning "over in a day." However, the adult stage of the mayfly is preceded by a larval stage which lasts considerably longer. It is not the *total* life of the mayfly that is ephemeral.

STEP 11

100 000 seconds (10^5 s)
1.16 days ($10^{0.06}$ d)

A hundred thousand seconds is equal to 27 7/9 hours, but here we must move once again to a larger unit of time.

In twenty-four hours, the Earth turns once upon its axis. This produces a period of day and night, probably the first time interval that impressed itself on the notice of primitive humanity.

The day and the night were each divided into 12 hours. This was convenient since 12 can be divided evenly by 2, 3, 4, and 6. Originally, as day and night altered their lengths with the seasons, the hours did the same. In the summer, therefore, the twelve hours of daylight were each longer than the twelve hours of nighttime, while, in the winter, the reverse was true. Then clocks improved, and it was quite realized that the change in duration for day and night cancelled each other, so that the duration of one day plus one night was equal at all times of the year, hours were given fixed lengths, regardless of whether the Sun was present in the sky or not. It follows then that in spring and summer, the daytime is longer than 12 hours and nighttime shorter (except at the equator), while in the fall and winter the reverse is true.

The 24-hour period of day plus night is now called "day," so that the word refers *both* to the 24-hour period, and to the time during which the Sun is above the horizon.

Actually, the 24-hour period is the "solar day." That is the time interval required to take the Sun from a particular position in the sky back to the same position: from noon to noon, for instance. (Actually, the time from noon to noon varies slightly with the time of the year because the Earth's orbit about the Sun is not quite circular, and because the Earth's axis is not perpendicular to the plane of revolution. However, 24 hours is the average time from noon to noon, and the deviations from that figure are not great. The "mean solar day" is exactly 24 hours, or 86 400 seconds long.)

The "sidereal day," or the time it takes for a given star to seem to circle the Earth and return to its original place, is a trifle shorter than the solar day (because the Sun seems to drift forward against the background of the stars, thanks to the motion of the Earth about the Sun). The sidereal day is 23.934 hours long, or 0.997 solar days long. The difference is about 3.93 minutes.

The sidereal day is of interest only to astronomers, however. To all others, it is the solar day, and only the solar day, that counts.

The solar day can be used in the SI version of the metric system, and is symbolized "d." The Step-11 time interval of 100 000 seconds is equal to 1.16 days.

The rotation period of Mars is a trifle longer than that of Earth. The sidereal day of Mars is 1.026 days long.

As for the various satellites of the Solar system, the three innermost sizable satellites of Saturn—Mimas, Enceladus, and Tethys—revolve about Saturn in periods, respectively, of 0.942 days, 1.370 days, and 1.888 days. Miranda, the innermost satellite of Uranus, revolves about it in 1.414 days. Io, the innermost of the Galilean satellites, revolves about Jupiter in 1.769 days. Deimos, the farther of the two Martian satellites, revolves about Mars in 1.262 days.

316 000 seconds ($10^{5.5}$ *s*)
3.66 days ($10^{0.56}$ *d*)

We are now approaching time units longer than a day. The shortest of these is the "week," which is defined as 7 days, and which is not used in the SI version of the metric system. (In fact, nothing, strictly speaking, is used in the SI version, except seconds, minutes, hours and days.)

The week represents the shortest time interval that is inspired by the Moon, which is second only to the Sun as a prominent object in the sky. The Moon goes through an unending and repetitive change of appearance or "phase" from night to night. Four of these are particularly noted: "new" when the Moon is so near the Sun in the sky that virtually none of the visible side receives sunlight, so that the Moon can be seen, if at all, as only a vanishingly thin crescent; "first quarter" when the lighted portion has expanded until what is visible is a semicircle of light; "full" when the lighted portion has spread over the entire visible side of the Moon, so that we see a perfect (or nearly perfect) circle of light; and "last quarter" when the lighted portion has shrunk to a semicircle again. The lighted portion continues to shrink until it is a new Moon again and the cycle starts over.

The average time lapse from one of these phases to the next is 7.38 days, so that if the Moon is at first quarter on Sunday, it will be full the next Sunday. The "week" thus marks out the important phases of the Moon, and the word is sometimes traced to an old Teutonic word meaning "change (of phase)." The German word for "change" is "Wechsel," and that seems rather similar to the German "Woche" meaning "week."

The change of phase does not occur every seven days exactly, however, and the length of time from full to new, or from first quarter to last quarter, is 14.76 days (or nearly 15 days). Thus, if it is new Moon on a Sunday, it will be full Moon not on the second Sunday thereafter, but on the Monday following it. To keep the week in good time with the phases of the Moon, it would be necessary to keep to a pattern of weeks in which some had seven days and some had eight.

Thus, you might collect the weeks into groups of eight, and within each group have the weeks consist of 7 days, 8 days, 7 days, 7 days, 8 days, 7 days, 8 days, 7 days, then start the pattern over with the next group of eight. In this case, the beginning of each week would correspond to one of the four important phases. (Not quite exactly, even so. There would have to be slight changes now and then.)

This was never done, perhaps because it was inconvenient, and perhaps because the ancient Sumerian astronomers noted that there were seven easily visible heavenly bodies that moved against the background of the stars: those known to us as Sun, Moon, Mercury, Venus, Mars, Jupiter, and Saturn. This apparently gave the number seven a mystic significance, and the week was kept at that number even though it meant that the phases of the Moon would not fit.

In any case, the Step-12 time-level of 3.66 days is very close to half a week long (0.523 weeks).

Europa, the smallest of the Galilean satellites, circles Jupiter in 3.551 days. Dione and Rhea, two of the satellites circling Saturn, have periods of revolution of 2.737 and 4.518 days, respectively. Ariel and Umbriel, two of the satellites circling Uranus, have periods of revolution of 2.520 and 4.144 days, respectively. Triton, the larger of Neptune's two satellites, has a period of revolution of 5.877 days.

Pluto, the farthest of the known planets, rotates on its axis once in 6.39 days.

STEP 13

1 000 000 seconds (10^6 s)
11.57 days ($10^{1.06}$ d)

At the Step-13 time interval, we are dealing with a period that is roughly 1 2/3 weeks long. The familiar two-week vacation is only slightly longer. A period of two weeks is sometimes called a "fortnight" (especially in Great Britain), which is an abbreviation of "fourteen nights." The Step-13 time interval is just about five sixths of a fortnight.

We are in the range of the total lifetime of organisms large enough to be seen. The male housefly lives about 17 days from egg to death, even if it has plenty to eat and escapes death by misadventure.

Oberon, the outermost known satellite of Uranus, circles its planet in 13.463 days; Titan, the largest of Saturn's satellites, circles Saturn in 15.945 days; and Callisto, the farthest of the Galileans, circles Jupiter in 16.689 days.

STEP 14

3 160 000 seconds ($10^{6.5}$ s)
36.57 days ($10^{1.56}$ d)

The period from new Moon to new Moon—that is, for one complete cycle of the phases of the Moon—is equal to 29.53 days. This is equal to the synodic period of revolution of the Moon about the Earth, and is called a "synodic month," where the word "month" clearly comes from "Moon."

The Babylonians developed a "lunar calendar" in which the first day of the month always came at the time of the new Moon. This was adopted by many other peoples, including the Jews and the Greeks. In order to keep the first day of the month at the new Moon, some months had to be 29 days long and some 30 days long according to a fixed and rather complicated system.

In modern times, we have abandoned the connection of the month and the Moon, and make uses of months that are irregular enough in length, one being as short as 28 days, and some as long as 31 days. The average length of these "calendar months," however, is 30.437 days. A calendar month is therefore equal to 1.03 synodic months.

As for the Step-14 time range of 36.57 days, that is equal to just about 1.2 calendar months.

The synodic month is the Moon's revolution about the Earth relative to the Sun. The "sidereal month" is the Moon's revolution about the Earth relative to the stars. The sidereal month is over two days shorter than the synodic month, for astronomical reasons that need not concern us. The sidereal revolution of the Moon is 27.322 days long, or just about 0.9 calendar months.

The shortest presidential administration in American history was that of William Henry Harrison who, in 1841, was our ninth president for 31 days, before dying of natural causes.

STEP 15

10 000 000 seconds (10^7 s)
115.7 days ($10^{2.06}$ d)

The Step-15 time duration is about 4 months long.

The innermost planet, Mercury, circles the Sun in 88 days (compare this with Saturn's satellite Iapetus, which circles its planet in 79.331 days).

The average life span of the human red blood corpuscle is about 125 days.

James Abram Garfield was twentieth president of the United States in 1881 for 199 days (6.5 months), before dying of the wound dealt him by an assassin.

STEP 16

31 600 000 seconds ($10^{7.5}$ s)
365.7 days ($10^{2.56}$ d)
1 year (10^0 y)

This time interval is remarkably close to the period of revolution of the Earth about the Sun, which is 365.2422 days. This is the "tropical year," and the Step-16 time interval is thus equal to 1.001 tropical years.

The tropical year is not a unit permitted in the SI version of the metric system, but it is so commonly and universally used, and would be so convenient to my purposes, that on this occasion I am going to rebel and use it, giving it the symbol of "y."

The tropical year, which is actually the period of time between successive vernal equinoxes, is very nearly 365 1/4 days long so that three successive years are 365 days long each, while the fourth, the "leap year," is 366 years long. Since the tropical year is not quite 365 1/4 days long, three years in every four hundred which would ordinarily be leap years are not. This arrangement, called the "Gregorian calendar," after Pope Gregory XIII, who first introduced it to Europe in 1582, works very well and is now accepted by virtually all nations—at least for international dealings, if for nothing else.

The period of revolution of the Earth about the Sun with respect to the stars is the "synodic year," which is 365.256 days long, or 20 minutes longer than the tropical year.

The length of the year reflects itself on Earth as the cycle of the seasons. The first human calendars were based on the lunar month, and it became apparent that the seasons repeated themselves every twelve lunar months.

Not precisely, however. Twelve lunar months come to a total of 354 days. This is the "lunar year," and it is equal to 0.97 tropical years. To put it another way, the tropical year is equal to 12.37 lunar months.

In order to keep the lunar months even with the tropical year, the lunar years were divided by the Babylonians into groups of 19, of which 12 were made up of 12 lunar months, and 7 of 13 lunar months, in a fixed pattern. This system was adopted by the Jews and Greeks.

The Egyptians preferred to use a different system whereby the tropical year was divided into 12 months, each of which was 30 days long, with 5 extra days added to keep the year even with the seasons (but they made no provision for a leap year). In the time of Julius Caesar, the Romans adopted it, with some modifications, and, with further modification by Pope Gregory, the Egyptian calendar is the one the world now uses.

Some satellites have periods of revolution roughly comparable to that of Earth. Neptune's outer satellite, Nereid, circles its planet in 0.985 years (359.88 days), while Saturn's outermost satellite, Phoebe, circles its planet in 1.51 years (550.4 days). The planet Mars circles the Sun in 1.88 years (687 days).

We are in the time range of the lifespan of the smallest mammals. The pygmy shrew, even if adequately fed and kept from harm, will not live for longer than a year.

Zachary Taylor was twelfth president of the United States in 1849 and 1850 for a period of 1.35 years (492 days), before dying of natural causes.

STEP 17

100 000 000 seconds (10^8 s)
3.16 years ($10^{0.5}$ y)

Members of the House of Representatives are elected for terms of 2 years.

At the Step-17 time duration, we have surpassed the periods of revolution of all known satellites. The longest is that of Sinope, Jupiter's outermost satellite, which revolves about its planet in 2.08 years (758 days), or just about the term of office of a member of the House of Representatives.

Comet Encke, the comet with the smallest known orbit about the Sun, has a period of revolution of 3.3 years. Ceres, the largest of the asteroids, circles the Sun in 4.6 years.

John Fitzgerald Kennedy, the thirty-fifth president of the United States, was in office for 2.84 years (1 037 days), at which time his life was cut short by an assassin's bullet. The normal length of a presidential term, as set by the Constitution, is 4 years. The first president whose administration was just that length was John Adams, the second president; the most recent 4-year president (at the time of writing) was James Earl (Jimmy) Carter, the thirty-ninth president.

The black rat has a maximum lifespan of a little over 3 years.

STEP 18

316 000 000 seconds ($10^{8.5}$ s)
10 years (10^1 y)

Various names have been given to periods of time longer than a year. For instance, there was a time when the Romans took some sort of census every five years, and called that a "lustrum," from a Latin word meaning "to wash," for there was a ritual purification of the public after the census. For that reason, a period of 5 years is sometimes referred to as a lustrum, but only in poetic or ornate writing styles.

Again, there were times in Roman history when tax assessments were revised and the results announced every fifteen years. This was called an "indiction" from a Latin word meaning "to announce," and the term is used (very rarely) for a period of 15 years.

The one common word for a period in this time range is "decade," from the Latin word for "ten." This represents the Step-18 time duration of 10 years.

American senators are elected for terms of 6 years or three fifths of a decade.

American presidents may be reelected, in which case, if no misfortune intervenes, they may serve for 8 years, or four fifths of a decade. The first to serve for 8 consecutive years was George Washington, the first president, and the most recent, so far, was Dwight David (Ike) Eisenhower, the thirty-fourth president. Grover Cleveland served 8 years, but these were not successive. The 4-year administration of Benjamin Harrison intervened, so that Cleveland was both the twenty-second and twenty-fourth president.

Prior to 1951, when the Twenty-second Amendment of the Constitution was adopted, the years served by an American president were not legally limited to 8 (though custom imposed such a limit). Franklin Delano Roosevelt, the thirty-second president, defied the custom, ran, and was elected to third and fourth terms. His fourth term was cut short by natural death, and his administration lasted 12.115 years, the longest in American history.

By an odd fatality, Adolf Hitler ruled Germany for a period of time that almost exactly matched the period during which F. D. Roosevelt presided over the United States. Hitler came to power 33 days before Roosevelt was inaugurated, and committed suicide 18 days after Roosevelt's death, so that his rule lasted 12.255 years.

The period of revolution of Jupiter about the Sun is a little shorter than the administration of F. D. Roosevelt, being 11.86 years.

The maximum lifespan of a marmoset is about 10 years.

STEP 19

1 000 000 000 seconds (10^9 *s*)
31.6 years ($10^{1.5}$ *y*)

The most important nonspecific unit of time is the "generation," a term frequently used to represent the time it takes for people to be replaced by their children. It is most frequently used for a period of 33 years, which puts it very nearly at the Step-19 time duration.

Justices of the Supreme Court are appointed for life (or until they voluntarily resign, or are impeached and convicted). As of the time of writing, the justice who served longest was William Orville Douglas, who served 36.54 years from 1939 to 1975.

Popes are elected for life (or until they resign or are deposed), and the longest papal reign is that of Pius IX, who reigned from 1846 to 1878 for 31.65 years—for just over a billion seconds.

Rulers who seize power and rule nondemocratically can sometimes remain in power for extended periods. Benito Mussolini ruled Italy for 21 years; Francisco Franco ruled Spain for 36 years; Antonio Salazar ruled Portugal for 42 years.

Kings and others who rule by hereditary right have on occasion done better still, for they sometimes succeed to the throne at an early age.

The living ruler of an important nation who has been on his throne longest is Hirohito, the emperor of Japan. He succeeded to the throne in 1926 at the age of 25, and has now been ruling 57 years, and is 82 years old.

Two English monarchs have reigned longer than that: George III, who ruled for 60 years from 1760 to 1820, having become king at the age of 22; and Victoria, who ruled for 64 years, from 1837 to 1901, having become queen at the age of 18. She reigned for just over two billion seconds.

In general, mammals live longer, the larger they are. A camel can live 27 years, a giraffe, 28, and a lion, 29. An Asian rhinoceros can live up to 32 years, or just over a billion seconds; an orangutan can live 34 years, a zebra, 40 years, a mandrill baboon, 46 years, a European brown bear, 47 years, and a hippopotamus, 54 years.

Among birds, various eagles can live into their 40s and 50s, and an Andean condor has been known to live 65 years. The greatest age recorded for a snake is that of a python, who lived 34 years. Alligators and lizards have reached their 50s. Giant salamanders, largest of the amphibians, also have reached their 50s. An American lobster may live to be 50 years old, too.

Saturn has a period of revolution about the Sun of 29.46 years. Sirius and its white-dwarf companion, Sirius B, circle each other within a period of 49.94 years.

There are atom-varieties, or isotopes, with half-lives of all ranges. Many of them are extremely unstable and have half-lives of only a few seconds, or minutes. They exist only because they have been formed out of much more nearly stable, naturally occurring isotopes, or because they have been formed by scientists through nuclear reactions in the laboratory.

Some isotopes have half-lives that are several decades long. Argon-42 has a half-life of 33 years, for instance; and strontium-90 has a half-life of 27.7 years. (Exploding nuclear bombs produce various radioactive isotopes, including strontium-90. Strontium-90 is chemically similar to the calcium in the bones, and the isotope would get into the bones by natural processes. Once in the bones, it would remain for extended periods, slowly breaking down and subjecting the body to dangerous radiation. It was, in part, the spectre of strontium-90 that, in 1963, persuaded the major nuclear powers, the United States, Great Britain, and the Soviet Union, to give up the practice of testing such nuclear bombs in the open air.)

STEP 20

3 160 000 000 seconds ($10^{9.5}$ s)
100 years (10^2 y)

A time interval of 100 years is known to all of us as a "century" from the Latin word for "hundred."

If we leave human beings out of account, no mammal, however large, can survive a century. The longest authentic age at death, for a nonhuman mammal, is 69 years for an Asiatic elephant. A few species of birds are known to live into their late 60s, and probably some may live into their 70s.

Some fish do even better. The longest-lived fish may be sturgeons, and one sturgeon was caught and estimated to be 82 years old from the growth rings on its fins. Some sea anemones have lived to be 90 years old, and these are thought to be the longest lived of all invertebrates.

And yet human beings can, and do, surpass the century mark, outliving all mammals without exception, even those tens and hundreds of times as massive. (Why this is so is not known—but let's not complain.) In fact, some monarchs have reigned respectable fractions of a century.

Francis Joseph I, emperor of Austria, reigned for 68 years, better than two thirds of a century, from 1848 to 1916, having come to the throne at the age of 18. Louis XIV of France sets the modern record with a reign of 72 years, nearly three fourths of a century, from 1643 to 1715, having come to the throne at the age of 5.

The all-time record for length of reign is thought to be that of Pepi II, of Egypt's Sixth Dynasty, who came to the throne at the age of 6, about 2566 B.C., and is supposed to have reigned for 90 years (!), dying at the age of 97.

Among centenarians is Bernard de Fontenelle, a French writer (1657–1757), who died just a few weeks before his hundredth birthday. William David Coolidge (1873–1974), who made possible the use of tungsten filaments in light bulbs, lived to be 101, and Michel Eugene Chevreul (1786–1889), who did important work on fats, lived to be 103.

The longest human life for which satisfactory proofs of birth and death exist are those of a French Canadian, Pierre Joubert, who was born on July 15, 1701, and died on November 16, 1814. He died at the age of 113.34 years, after living more than 3.5 billion seconds. His heart undoubtedly beat 4 billion times before stopping, whereas nonhuman mammals, even under the most favorable conditions, rarely endure for more than 1 billion heartbeats. It seems quite certain that when Joubert died, not a single mammal anywhere in the world (human or nonhuman), was still alive who had been alive when he had been born.

The only animals of any kind that are definitely known to outdo the human being in longevity are some tortoises. The greatest age reliably reported for a tortoise at the time of death was 152 years (or nearly 5 billion seconds).

Astronomically, the most famous of all comets, Comet Halley, has a period of revolution about the Sun of 76.1 years. The planet Uranus has one of 84.0 years, and Neptune, one of 164.8 years. (It may be that no Earthly animal can survive a single turn of Neptune about the Sun.)

The two stars of the Alpha Centauri binary system revolve about a common center of gravity in 80.09 years, while the two stars of the Gamma Virginis system do so in a period of 171.4 years.

Of the isotopes, molybdenum-93 has a half-life of at least 100 years, and terbium-157 has one of 150 years.

We have reached the stage now where we can look back in history over the time interval we are dealing with and note considerable change. A hundred years ago, there were no computers, no nuclear bombs, no television, no radio, no airplanes, no movies, no automobiles. Chester Alan Arthur was president of the United States, and Morrison R. White was chief justice. Queen Victoria was about to celebrate her golden jubilee after fifty years on the throne of Great Britain; William I was the kaiser of Germany and approaching the end of his long life; and Leo XIII was pope.

STEP 21

10 000 000 000 seconds (10^{10} s)
316 years ($10^{2.5}$ y)

At the Step-21 time-level, we are well beyond the realm of animal longevity, but the plant kingdom remains. There are many types of trees that can live for more than three centuries.

And though human individuals are left behind, human institutions can outlive the human beings who founded them. The United States (counting from the Declaration of Independence on July 4, 1776) has been in existence for 207 years, at this time of writing. Individual states are older still. The youngest of the thirteen original states is Georgia, founded in 1733, so that it is 250 years old.

If we look back 316 years ago, neither the United States nor the state of Georgia was in existence, but Englishmen had arrived in what is now Massachusetts 46 years before, and in Virginia 59 years before. The city of Quebec was 58 years old, and Canada was French. New Amsterdam was 43 years old and had just been taken by the English two years before and renamed New York.

Charles II was King of England, having been restored to the throne six years before, and London was suffering a plague and a great fire. Newton was a young man, discovering how white light could be broken up into colors, and already working out the law of gravity. Alexander VII was pope; Louis XIV of France was still a young man. Alexis was tsar of Russia, Leopold I was Holy Roman emperor, and Charles II had just become king of Spain. The Wars of Religion were finally over, and the Ottoman Empire was at the peak of its territorial expansion.

Pluto, the most distant of the planets, revolves about the Sun in 247.1 years, and the two stars of the Eta Cassipeiae binary system revolve about a common center of gravity in 401 years.

The half-life of argon-39 is 269 years, and that of niobium-92, about 350 years.

STEP 22

31 600 000 000 seconds ($10^{10.5}$ s)
1 000 years (10^3 y)

A thousand years is commonly called a "millennium" from Latin words meaning "a thousand years." Even this length of time does not outstretch the capacity of a tree to endure. The average lifespan of a redwood tree is about 1 100 years.

A thousand years ago, England was under Edgar and was still Anglo-Saxon. It was not until 83 years later that William of Normandy was to defeat the Saxons at Hastings and become William the Conqueror. France was under Lothair, a descendant of Charlemagne. It was not till four years later that Hugh Capet was to become king of France and found the modern line of French kings. Otto II was Holy Roman emperor, and Basil II was Byzantine emperor. (The Byzantine Empire was then at a peak of military strength.)

The isotope terbium-158 has a half-life of 1 200 years. Radium-226, the longest-lived radium isotope, has a half-life of 1 602 years.

STEP 23

100 000 000 000 seconds (10^{11} s)
3 160 years ($10^{3.5}$ y)

Now we approach the best that any living organism can do in the way of longevity. The giant sequoia trees, the most massive of all living organisms, are among the very few species of organisms that can live for over a hundred billion seconds. Some that are still alive may be 3 750 years old (33 times the maximum lifespan of a human being).

A bristlecone pine growing in eastern Nevada would seem to be the oldest living thing still alive as far as we can tell. Its age is estimated at 4 900 years (though another tree in the grove which, unfortunately, had to be cut down, seems to have been just over 5000 years old (45 times the maximum lifespan of a human being).

The half-life of carbon-14 is 5 600 years, which makes it ideal for dating old pieces of organic matter—wood, charcoal, textiles, and so on. Much of the dating of ancient times and prehistory now depends upon carbon-14 analysis.

If we look back in history a hundred billion seconds, we find that 3 160 years ago (1177 B.C.) the Israelites were in the period of the Judges, and Gideon was fighting his battles against the Midianites.

The Egyptians under Rameses III were fighting against "The Peoples of the Sea" and had come to the end of their great days. The Assyrians were laboring to build an empire.

The Great Pyramid, constructed about 2500 B.C., was already 14 centuries old in 1178 B.C. and has now endured for 4 500 years, yet it is not quite as old as a still-living bristlecone pine. It is rather an exciting thought that a frail piece of life exists on this earth that is older than the pyramids.

STEP 24

316 000 000 000 seconds ($10^{11.5}$ s)
10 000 years (10^4 y)

Step 24 takes us beyond the limits of history. The details which go to make up history can only exist where written matter has survived to give us names and events as known to contemporary witnesses.

Writing was first invented in Sumeria, about 3200 B.C., so that history has an extreme duration of only 5 200 years, and is only three centuries older than the oldest living bristlecone pine.

Civilization is, however, older than history. We can date civilization (which is from a Latin word meaning "city dweller") from the first foundation of cities. The oldest city in the Middle East may have been founded as long ago as 9000 B.C. Civilization would, then, be just about 11 000 years old.

A literal interpretation of the early books of the Bible would lead one to believe that the Earth and the entire Universe were not much older than the existence of writing. Most editions of the King James Bible give the date of the divine creation as 4004 B.C., in accordance with a calculation by James Ussher, a bishop of the Anglican Church, a little over three centuries ago. That would make the Universe just about 6 000 years old.

Suppose the old bristlecone pine began as a cone produced by a parent pine that was already a little over a thousand years old at the time. In that case, if James Ussher were correct, the old bristlecone pine's parent tree might have been created by God and the one now alive would be only the second generation.

According to medieval Jewish calculations, the Universe was created in 3761 B.C., which would place its present age at only a little over 5 740 years. The Jewish religious calendar numbers the years from this supposed date of creation. Some Greek Orthodox calculations place the creation as long ago as 5509 B.C., and make the present age of the Universe nearly 7 500 years.

Contemporary creationists, who cling to the literal words of the Bible, are willing to speak of the Universe as fully 10 000 years old.

Even this, however, would not be as old as the oldest cities, and since science does not accept creationist legends (and, on the basis of unarguable evidence, cannot do so), we will proceed up the ladder of time.

STEP 25

1 000 000 000 000 seconds (10^{12} s)
31 600 years ($10^{4.5}$ y)

The axis of the Earth lies in a nearly fixed direction that is at an angle of about 23.5 degrees from the perpendicular to the plane in which it revolves about the Sun. It is this tilting of the axis that causes the Sun to move higher and lower in the sky as the Earth revolves about the Sun, and that causes the changes of the seasons and the alternation of winter and summer.

The axis, while maintaining its angle of tilt (with very slow and trifling variations about an average) marks out a slow circle. The northern end of the axis, if extended to the sky, intersects the sky very near the position of the North Star. As the years pass, that point of intersection will slowly skim by the North Star, move away, mark out a circle, and eventually return to the North Star. This motion is referred to as "the precession of the equinoxes."

The axis marks out a complete circle in 25 784 years. In other words, when the axis last marked out the North Star, human beings had not yet become civilized anywhere in the world. And when the axis next marks out the North Star, what then?

About 31 600 years ago, the Earth was deep in its most recent ice age, and human beings, while uncivilized, were producing magnificent artistic works—painting in caves, and fashioning tiny tools made of bone and ivory, together with larger tools made of flint. In the 31 600 years that have elapsed since then, a ray of light could travel from here to the center of the Galaxy.

Protactinium-231 has a half-life of 32 500 years, very nearly the Step-25 time-level.

STEP 26

3 160 000 000 000 seconds ($10^{12.5}$ s)
100 000 years (10^5 y)

If we look back 100 000 years, then the most recent ice age (the fourth in this geologic period, according to the usual count) had not yet arrived. The world was in a mild "interglacial period," as it is now. "Modern man" was not in evidence. *Homo sapiens* existed even that long ago, however, but in another variety usually spoken of as "Neanderthal man."

Thorium-230 has a half-life of 80 000 years.

STEP 27

10 000 000 000 000 seconds (10^{13} s)
316 000 years ($10^{5.5}$ y)

The largest orbits in the Solar system are those of the comets. Comet Kohoutek, which made its closest approach to the Sun toward the end of 1973, has such an enormous orbit that it takes 216 500 years to complete one turn about the Sun.

About 316 000 years ago, *Homo sapiens* did not exist. There were, however, other hominids (organisms more closely related to human beings than to apes) in existence. These were *Homo erectus,* more familiarly known as "Java Man" and "Pekin Man." Even at that time, these hominids may have been advanced enough to have discovered the use of fire.

Technetium-99, the most long-lived technetium isotope, has a half-life of 212 000 years.

STEP 28

31 600 000 000 000 seconds ($10^{13.5}$ s)
1 000 000 years (10^{6} y)

There are periods of revolution even at this stage. Proxima Centauri, the dim red dwarf that is the closest star to ourselves, may revolve about the center of gravity of the two large stars of the Alpha Centauri system in a period of about 1 250 000 years. The distant comets of our own Solar system may have periods just as long or longer.

If we look back a million years into Earth's history, we find we are at the beginning of the "Pleistocene Era," the most recent of the geologic eras. The ice ages have not yet begun, but *Homo erectus* already exists, although it has probably not yet discovered the use of fire. (Aluminum-26 has a half-life of 740 000 years; manganese-53 has one of 1 900 000 years.)

STEP 29

100 000 000 000 000 seconds (10^{14} s)
3 160 000 years ($10^{6.5}$ y)

Three million years ago, *Homo erectus* did not yet exist. The hominids that existed were an earlier, smaller group whose traces were first found in southern Africa and are called *Australopithecus* ("southern ape," although they were closer to human beings than to apes). They first evolved about 4 000 000 years ago. Earth was in the "Pliocene Era." (Cesium-135 has a half-life of 3 000 000 years.)

STEP 30

316 000 000 000 000 seconds ($10^{14.5}$ s)
10 000 000 years (10^{7} y)

Ten million years ago, the earliest hominid we know of, *Ramapithecus,* was spreading into Asia, after having evolved in east Africa about twenty million years ago. (Iodine-129 has a half-life of 17 000 000 years.)

STEP 31

1 000 000 000 000 000 seconds (10^{15} s)
31 600 000 years ($10^{7.5}$ y)

Thirty million years ago, the Pliocene Era had not yet begun, and Earth was in the middle of the era before that, the Miocene. The first hominids had not yet appeared, but apelike organisms were evolving —such as *Dryopithecus,* the common ancestor of the gorilla, the chimpanzee, and the hominids. The first apes appeared about forty million years ago. (Lead-205 has a half-life of 30 000 000 years.)

STEP 32

3 160 000 000 000 000 seconds ($10^{15.5}$ s)
100 000 000 years (10^{8} y)

If we go back in time 100 000 000 years, we find ourselves in the Cretaceous Era. The primates (the group to which all the lemurs, monkeys, apes, and hominids belong) are just evolving, and have reached the primitive lemur stage. (Monkeys did not evolve until 70 000 000 years ago.)

All mammals were small and insignificant at this time, and the dominant species on Earth were the large reptiles popularly called "dinosaurs." They disappeared rather suddenly at just about the time that monkeys were evolving. (Plutonium-244 has a half-life of about 76 000 000 years.)

Some stars do not remain in the stable, steadily shining stage (the main sequence) for very long. The more massive the star, the shorter its stay on the main sequence. Stars of spectral class B (which are from three to nine times as massive as our Sun) will remain on the main sequence for not more than 100 000 000 years, and will then explode, collapse, or both. Any star of spectral class B, shining in the sky today, we can safely suppose to have formed no earlier than the time when the dinosaurs were completing their sojourn on Earth.

STEP 33

10 000 000 000 000 000 seconds (10^{16} s)
316 000 000 years ($10^{8.5}$ y)

Three hundred million years ago, there were no mammals; and no birds, either. Even the dinosaurs had not yet evolved; but their primitive reptilian ancestors were coming into being.

The land at that time had not been long colonized by life. The first land animals appeared about 380 000 000 years ago, and the first land plants appeared about 425 000 000 years ago. Before that, Earth's land surface was sterile, although, of course, there was life in the sea.

The first fossils of any kind, even of sea life, are about 600 000 000 years old.

Uranium-235 has a half-life of 710 000 000 years. Isotopes with half-lives shorter than this would not be of primordial origin. In other words, if substantial quantities of a particular isotope had existed on Earth at the time of its formation, none but traces would exist now if it had had a shorter half-life than uranium-235. As it is, since half of any uranium-235 disappears after 710 000 000 years, half of the remaining half after another 710 000 000 years and so on, only about 1/100 of the uranium-235 present at the time of the Earth's formation still exists. That is enough, however, to allow substantial amounts of it to remain. One out of every 140 uranium atoms in the crust is uranium-235.

STEP 34

31 600 000 000 000 000 seconds ($10^{16.5}$ s)
1 000 000 000 years (10^9 y)
1 eon (10^0 e)

A billion years is sometimes said to be equal to 1 "eon" (which I will symbolize as "e" and will, for convenience' sake, use).

Stars of spectral class A, including Sirius, Vega and Altair, for instance, which are 2 or 3 times as massive as the Sun, have a lifetime on the main sequence of 0.5 to 1 eon. Any such star, shining in the sky today, must have been formed at a time when life had already existed in Earth's seas for some eons.

Nevertheless, the life existing in Earth's seas an eon ago existed only in the form of simple invertebrate organisms, perhaps little better than microscopic in size at most.

Potassium, as found in nature, is made up of three isotopes. Two of these, potassium-39 and potassium-41, are stable and make up virtually all of that element there is. However, there is also potassium-40, which is slightly radioactive, and which has a half-life of 1.26 eons (1.77 times that of uranium-235). This means that only about 1/13 the primordial supply of potassium-40, the amount present at the time of Earth's formation, still exists. Even so, 1 potassium atom out of every 850 in existence is potassium-40.

All living tissue contains potassium, which is an element essential to life. Our own bodies contain many trillion potassium atoms, and 1 out of every 850 of these is radioactive potassium-40. This is not very alarming, however. The longer the half-life of a radioactive atom, the fewer the breakdowns per second, and by the time we reach half-lives in the eons, the number of breakdowns are too few to cause us great worry.

STEP 35

100 000 000 000 000 000 seconds (10^{17} s)
3.16 eons ($10^{0.5}$ e)

Three eons ago, life on Earth had just evolved, and it existed in primitive forms no more complicated than bacteria or the simplest algae, if that. The oldest rocks that have been successfully dated on Earth are about 3.76 eons old, but the Earth has to be older than that. As a matter of fact, a variety of lines of evidence point to the Earth, and, indeed, the entire Solar system, as having been formed about 4.6 eons ago.

The half-life of uranium-238, the stablest isotopic form of that element, is about 4.51 eons. This means that half the primordial supply of uranium-238 still exists today. Since uranium-238 has a half-life that is 6 1/3 times that of uranium-235, so that uranium-238 breaks down much more slowly than uranium-235 does, the former constantly gains on the latter. As the eons pass, uranium becomes more and more heavily uranium-238. At the present time there are 556 atoms of uranium-238 for every 4 atoms of uranium-235. At the time the Earth was formed, there was 2.7 times as much total uranium of both forms as there is now, and there were only 11 atoms of uranium-238 for every 4 of uranium-235.

A star like Procyon, of spectral class F and only 1.25 times as massive as the Sun, would have a total lifetime on the main sequence of about 4 eons. It must be somewhat younger than the Sun at best, therefore, or it would be on the edge of expanding into a red giant (and then collapsing) by now.

STEP 36

316 000 000 000 000 000 seconds ($10^{17.5}$ *s*)
10 eons (10^{1} *e*)

Ten eons ago, there was no Earth, no Sun, no Solar system; only a vast cloud of dust and gas where, more than five eons later, the Solar system would form. There were, however, other stars (with accompanying planetary systems, no doubt) that already existed. Some of those stars still exist, while others have long since come to an end of their stay on the main sequence and have expanded, then collapsed.

The stars that existed ten eons ago were collected into galaxies, which in turn formed clusters, and the whole made up the Universe. The Universe of ten eons ago, however, was much less voluminous, and its average density was much higher, than is true of the Universe today. Since the Universe is expanding steadily as time goes on, we would view it contracting if we look backward in time. Something like 10 eons in the past, the Universe may have had only one tenth the volume it has today, and though it still contained all its galaxies, these must have been crowded closer together.

If we go still farther back into the past, the Universe would continue to shrink down to a point. From that point, it formed in a vast explosion popularly known as the "big bang." The big bang took place perhaps 15 eons ago, and that is usually taken as the age of the Universe.

The half-life of thorium-232 (the only form of the element that occurs in nature) is 14.1 eons. This means that of all the thorium that existed in the Universe when the elements formed not long after the big bang, about half still exists today despite the continuing breakdown of thorium atoms through the entire vast lifetime of the Universe.

Nor does the age of the Universe limit our ascent of the ladder of time, for there exists time in the future as well as in the past. For instance, the Sun has been on the main sequence since soon after its formation, and it will continue to remain on the main sequence for many additional eons. At its level of mass, it may be that it will stay on the main sequence for as much as 12 eons altogether. Since the Sun has now existed in its present form for 4.6 eons, only three eighths of its main-sequence lifetime has elapsed. It is a sedate middle-aged star.

STEP 37

1 000 000 000 000 000 000 seconds (10^{18} s)
31.6 eons ($10^{1.5}$ e)

Stars that are somewhat less massive than the Sun, such as Alpha Centauri B or 61 Cygni A, have main-sequence lifetimes of up to 30 eons.

The element rubidium occurs in nature as two isotopes, rubidium-85 and rubidium-87, which are found in the ratio of 13 atoms of the former for every 5 atoms of the latter. Rubidium-85 is stable, but rubidium-87 is slightly radioactive, with a half-life of 48 eons. In the entire lifetime of the Universe, only about one eighth of the rubidium-87 atoms that originally existed have broken down.

STEP 38

3 160 000 000 000 000 000 seconds ($10^{18.5}$ s)
100 eons (10^{2} e)

Only the smaller red dwarfs can endure on the main sequence for Step-38 time periods. Stars like Barnard's star and Proxima Centauri may continue to dribble out their feeble bits of radiation for a total of 200 eons.

The half-life of samarium-147 is about 105 eons.

Beyond this step, we can point to longer durations of time only in connection with the half-lives of very mildly radioactive atoms, atoms so mildly radioactive that their half-lives are in the thousands and millions of eons. Naturally, the longer the half-life, the more difficult it is to detect the very few breakdowns taking place and to measure the actual value of the half-life. Suppose, for instance, we skip a number of steps—

. . .

STEP 57

10 000 000 000 000 000 000 000 000 000 seconds (10^{28} s)
316 000 000 000 eons ($10^{11.5}$ e)

This represents a time of 316 billion eons, over 2 billion times the total lifetime of the Universe, yet it is possible that molybdenum-130 has a half-life of about twice this length of time.

In view of this, is it possible that every atom would break down if we were to wait long enough? Apparently, all the atoms have been built up out of hydrogen-1 by processes that occurred first in the big bang and then at the core of stars. The nucleus of hydrogen-1 is simply a proton. Will, then, all atoms break down into individual protons if we wait long enough?

Is it possible that even protons are not stable, but break down—after incredible lengths of time—into less massive particles and that, eventually, the Universe will consist only of electrons, neutrinos, photons, and gravitons? Let us skip additional steps—

. . .

STEP 78

316 000 000 000 000 000 000 000 000 000 000 000 000 seconds ($10^{38.5}$ s)
10 000 000 000 000 000 000 000 eons (10^{22} e)

This represents a time of 10 billion trillion eons. According to current theories, it is just possible that this enormous length of time represents the half-life of the proton. As I write, physicists are attempting to arrange experiments in which a very occasional proton breakdown due to this instability can be measured.

The Step-78 time range is, so far, the longest period of time that has possible physical significance and, with that, we come to the end of the ascent of the ladder of time.

We must now return to the second and begin a descent.

The Ladder of Time
DOWNWARD

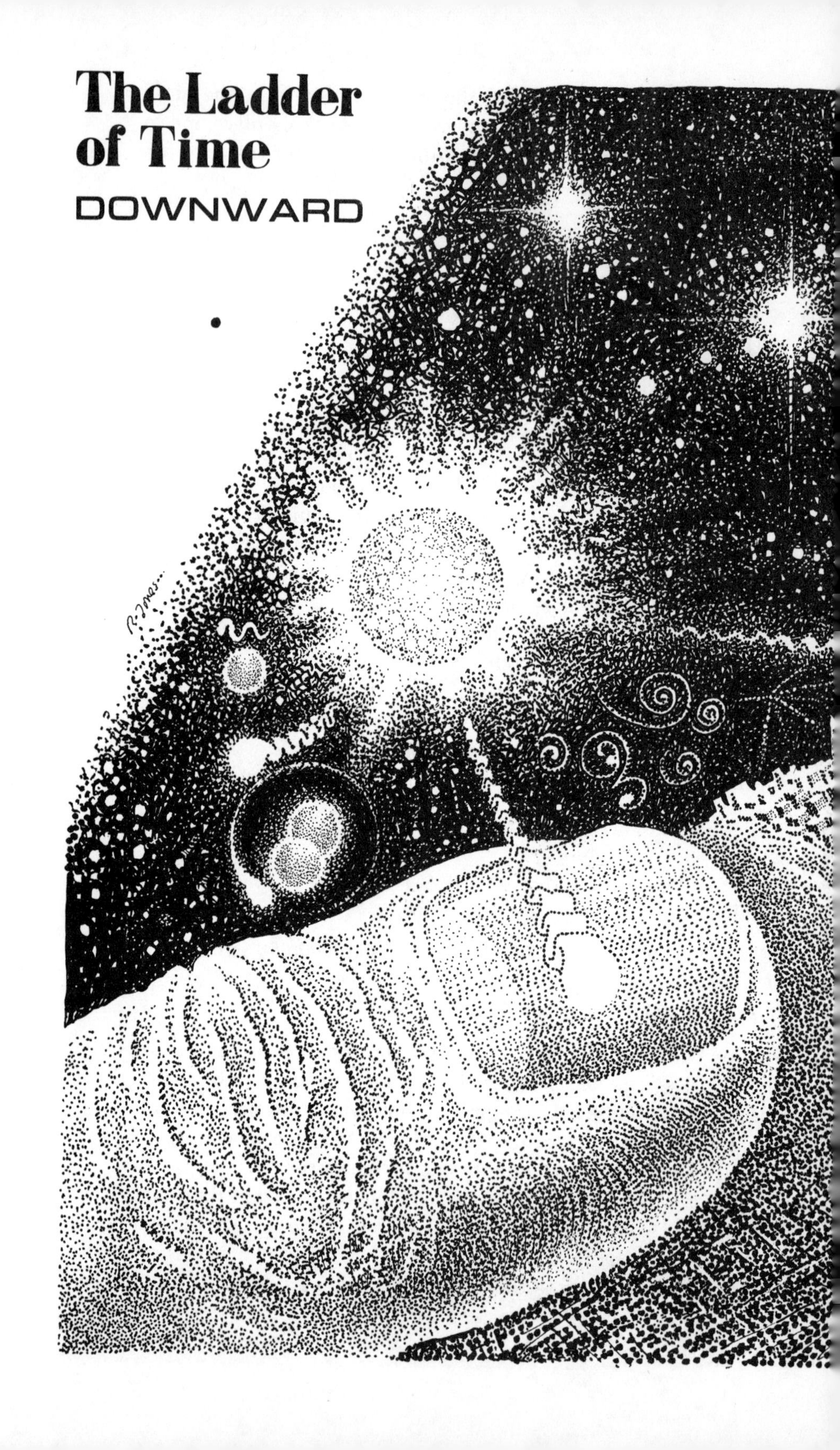

DOWNWARD

STEP 1

1 second (10^0 s)

The two other basic units of measure with which we have dealt, the metre and the kilogram, while not unduly large, have not seemed unduly small, either. Considerable fractions of either, a centimetre, for instance, or a gram, have seemed quite within the common grasp, so that the beginning, at least, of the descent of the ladder deals with familiar things.

The second of time, however, seems like a very small unit, and, indeed, it is the smallest with which we deal in ordinary life. Any fraction of a second would seem too small to deal with, and yet—a great deal can be done in a second.

The usual speed limit for automobiles in the United States, since the gasoline shortage of 1973, has been 88.5 kilometres (55 miles) per hour. In one second, an automobile moving at that speed limit will travel 2.46 dekametres (about three tenths of a city block).

In that same second, a sound wave (under ordinary Earth-surface conditions) will travel 3.32 hectometres through the air 4 1/8 city blocks). Earth will, in that second, travel 29.8 kilometres in its journey around the Sun, and light will travel 300 megametres (or, more precisely, 299.792 megametres).

With that in mind, we can begin the descent with greater confidence.

STEP 2

0.316 seconds ($10^{-0.5}$ *s)*
3.16 deciseconds ($10^{0.5}$ *ds)*

The prefixes of the metric system are commonly used for time intervals of less than a second. The prefixes "deci-" or "centi-" are almost never used, but they are permitted by the SI system, and I will use them.

In this interval of time, the wings of a small hummingbird will beat about 22 times, and there is a South American bird called the horned sungem that has wings that will beat 28 times in that interval. A tuning fork that sounds middle-C will vibrate 83.4 times.

Sound will travel 1.05 hectometres in this time, or 1.3 city blocks. Light, on the other hand, will travel 95 megametres, or 2 1/3 times around the Earth at the equator.

Silicon-25 has a half-life of 2.3 deciseconds; astatine-212, one of 3.0 deciseconds; and sodium-20, one of 3.9 deciseconds.

STEP 3

0.1 seconds (10^{-1} *s)*
1 decisecond (10^{0} *ds)*

A sound wave, in 1 decisecond, will travel 3.32 dekametres; while a supersonic jetplane, moving much faster than sound, can move as much as 9 dekametres, or 1.1 city blocks.

In that same decisecond, a satellite in low orbit about the Earth would move 8 hectometres, or 10 city blocks; while the Earth, in its motion about the Sun, would move 3 kilometres, or 37 1/2 city blocks.

A ray of light would, in a decisecond, travel 30 megametres. This is three fourths of the distance around the Earth at the equator, or nearly eight times the distance (as the plane flies) from New York to San Francisco.

The longest radio waves used in ordinary AM broadcasting will vibrate 55 000 times in a tenth of a second.

Helium-8 has a half-life of 1.22 deciseconds; carbon-9, one of 1.27 deciseconds; and lithium-9, one of 1.76 deciseconds.

STEP 4

0.031 6 seconds ($10^{-1.5}$ *s)*
3.16 centiseconds ($10^{0.5}$ *cs)*

There is a constant turnover of red blood corpuscles in the human body, and, in a period of 3.16 centiseconds, the number that come to an end of their useful life, and disintegrate, is nearly 73 000. This seems a frighteningly large number of red blood corpuscles to break down in so short a time, but there are 25 000 000 000 000 red blood corpuscles in the bloodstream of the average male adult human being, so 73 000 is a trifle in comparison. The human body is perfectly capable of producing 73 000 replacements in that same 3.16 centiseconds.

As an example of another sort of ending, 1 gram of uranium will experience 420 radioactive breakdowns in 3.16 centiseconds, while 1 gram of the far more unstable radium will, in the same interval of time, experience the breakdown of 1 170 000 000 radium atoms. Again, to undergo the loss of over a billion atoms in so short a time would make it seem that the radium will not take very long to vanish completely. However, a gram of radium contains 2 670 000 000 000 000 000 000, or over two trillion times as many atoms as manage to break down in this Step-4 time interval. It is not surprising, then, that it takes 1 620 years for half the radium to break down, and for uranium to endure correspondingly longer—especially as the number of atoms breaking down in each given unit of time declines as the total number of atoms remaining intact itself declines.

Curiously enough, we still have a conventional astronomic item to take note of. Tiny neutron stars, packing the ordinary mass of a star into a sphere that is only a few kilometres in diameter, rotate on their axes in very short periods. The neutron star in the Crab Nebula rotates once in 3.3099 centiseconds.

STEP 5

0.01 seconds (10^{-2} s)
1 centisecond (10^{0} cs)

In a centisecond, the human nerve impulse will move about 1 metre, or, roughly, from the brain to the hand.

The fastest wingbeat of a bird, that of the horned sungem, will produce not quite 1 beat in a centisecond. A honeybee's wings will beat 2 times, while those of a very small midge will beat 10 times. This last is the fastest known wingbeat of any organism.

In a centisecond, sound will travel 3.3 metres, about the width of the living room in a middle-class dwelling; but a light ray will travel 3 megametres, which is about the distance from Chicago to San Francisco.

STEP 6

0.003 16 seconds ($10^{-2.5}$ s)
3.16 milliseconds ($10^{0.5}$ ms)

The wings of a housefly will beat one stroke in this time interval; that of the midge with the fastest wingbeat, three strokes.

In 3.16 milliseconds, a rifle bullet will travel 1 1/4 metres and, in so doing, will be traveling faster than sound; for in that same time, a sound wave will travel only 1 1/8 metres, even while a tuning fork sounding middle-C is experiencing five sixths of a vibration.

A light ray will, in this interval, travel 948 kilometres, or the distance from Chicago to Washington, D.C., while the longest radio wave used for AM broadcasting is vibrating 1 740 times.

STEP 7

0.001 seconds (10^{-3} s)
1 millisecond (10^{0} ms)

In a millisecond, the human nerve impulse will travel about 1 decimetre, or about the width of an adult male hand, while the tiny midge with the fastest wingbeat will be able to manage a single beat.

A light ray will travel 300 kilometres, which is almost the distance from New York to Washington, D.C.

STEP 8

0.000 316 seconds ($10^{-3.5}$ s)
316 microseconds ($10^{2.5}$ μs)

At the Step-8 time interval, the world of man is beginning to freeze into motionlessness. An automobile, traveling at the usual speed limit, will only move 7.8 millimetres (about the length of the smallest fingernail), while even the fastest jet plane moves only 2.8 decimetres (about the length of the human foot).

A light ray, however, will travel 95 kilometres, or about the distance from New York to Trenton.

STEP 9

0.000 1 seconds (10^{-4} s)
100 microseconds (10^{2} μs)

In 100 microseconds, familiar astronomical bodies are no longer moving very much. In 100 microseconds, the Moon, in its orbit about the Earth, advances only 1.02 decimetres (the width of a hand); while the Earth, in its journey about the Sun, advances only 1.85 metres (about the distance from a man's feet to his head). Even light will travel only 30 kilometres, or just a little over the full length of Manhattan Island.

In 100 microseconds, just about 1 radioactive breakdown will take place in 1 gram of uranium, but 2 800 000 will take place in 1 gram of radium.

To take up a radically different subject, it was about 100 microseconds after the big bang that protons and neutrons formed. Earlier than 100 microseconds after the big bang, the then-tiny Universe consisted of quarks, particles even more fundamental than protons and neutrons.

STEP 10

0.000 031 6 seconds ($10^{-4.5}$ s)
31.6 microseconds ($10^{1.5}$ μs)

In 31.6 microseconds, Earth will move only 6 decimetres (about the distance from the knee to the sole of the foot) in its journey around the Sun. Light will travel 9.5 kilometres, or two fifths the length of Manhattan.

STEP 11

0.000 01 seconds (10^{-5} s)
10 microseconds (10^{1} μs)

In 10 microseconds, Earth will move only 1.8 decimetres (about the length of a hand) in its journey about the Sun, and light will travel 3 kilometres, or about three fourths the length of Central Park.

STEP 12

0.000 003 16 seconds ($10^{-5.5}$ s)
3.16 microseconds ($10^{0.5}$ μs)

Now we can get back to lifetimes, but not so much those of radioactive atoms, as those of subatomic particles.

Some subatomic particles are, as far as we know, perfectly stable and endure forever, if they do not encounter and interact with other particles. Examples are the electron and its opposite number (or antiparticle), the positron.

Other examples of stable particles are the three different neutrinos that are now known, and the corresponding three antineutrinos. (There is some possibility that neutrinos may "oscillate" from one variety to another, but even so, left to themselves, they remain neutrinos forever.)

Then, too, there is the photon (the fundamental particle of electromagnetic radiation, notably light), and the graviton (the fundamental particle, as yet undetected, of gravitational radiation). These have no antiparticles.

Then there are the particles of the atomic nucleus, the proton and the neutron. The proton (and antiproton) are perhaps not perfectly stable by current theories, but, if so, their half-lives are simply enormous.

The neutron (and antineutron) are unstable in isolation and have a half-life in the range of minutes, as was mentioned earlier in the book.

All other subatomic particles that have been detected are evanescent, and have lifetimes in the Step-12 range, or shorter.

Thus, there is a particle known as the muon (and antimuon) which has all the properties of an electron but is 208 times the electron's mass. The muon (like other short-lived particles) only exists when it is formed by energetic collisions that manage to transform some energy into mass in accordance with the relationships worked out in the theory of relativity. Once formed, such particles do not last long. The muon quickly breaks down to an electron and two different neutrinos, releasing much of the energy that had gone into its formation.

The lifetime of a muon (or antimuon) is only 2.2 microseconds, during which time light travels only 6.6 hectometres, or about four fifths the width of Central Park. During the lifetime of a muon, the longest waves used in AM radio broadcasting vibrate 1.22 times, while the longest microwaves vibrate about 700 times.

STEP 13

0.000 001 seconds (10^{-6} s)
1 microsecond (10^{0} μs)

In 1 microsecond—a millionth of a second—Earth will advance in its orbit about the Sun only 18.5 millimetres, the length of the end joint of the little finger. Light will travel 300 metres, or about 3 3/4 city blocks.

And in one microsecond, somewhere in the body, 2 red blood corpuscles will break down.

STEP 14

0.000 000 316 seconds ($10^{-6.5}$ s)
316 nanoseconds ($10^{2.5}$ ns)

In 316 nanoseconds, light will travel 9.5 dekametres, or just about 1 1/6 city blocks.

STEP 15

0.000 000 1 seconds (10^{-7} s)
100 nanoseconds (10^{2} ns)

In 100 nanoseconds, light will travel about 3 dekametres, or from one end of a typical suburban house to the other, and back.

STEP 16

0.000 000 031 6 seconds ($10^{-7.5}$ s)
31.6 nanoseconds ($10^{1.5}$ ns)

There are subatomic particles which are, for the most part, more massive than the muon and less massive than the proton. These are called "mesons" and are all, without exception, unstable and very short lived. Certain "K-mesons," or "kaons," have lifetimes of as much as 57 nanoseconds, about 1/40 that of a muon. (All the mesons, in the course of their rapid breakdowns, are eventually converted to electrons and neutrinos—or their antiparticles.)

In the course of the lifetime of this relatively long-lived kaon, light will travel about 1.7 dekametres, three times the length of an ordinary living room.

STEP 17

0.000 000 01 seconds (10^{-8} s)
10 nanoseconds (10^{1} ns)

The best-known mesons are the "pi-mesons" (or pions). The positive pion (and its antiparticle, the negative pion) have lifetimes of 26.1 nanoseconds, about 1/85 the lifetime of a muon. In the lifetime of a pion, light will travel 7.8 metres, or about the length of a long anaconda snake.

The best-known kaon, the positive kaon (and its antiparticle, the negative kaon) have lifetimes of 12.4 nanoseconds, during which light travels 3.72 metres, or about the width of an ordinary living room.

STEP 18

0.000 000 003 16 seconds ($10^{-8.5}$ s)
3.16 nanoseconds ($10^{0.5}$ ns)

In 3.16 nanoseconds, light will travel 9.48 decimetres, or about the length of the leg of an adult male human being. In this same interval of time, a gram of radium will experience 117 radioactive breakdowns, the longest microwaves will vibrate once, and the longest infrared waves will vibrate 7 600 times.

STEP 19

0.000 000 001 seconds (10^{-9} s)
1 nanosecond (10^{0} ns)

A nanosecond is a billionth of a second and, in that time, light will travel 3 decimetres, or about the average length of the foot of a male adult human being. The longest infrared waves will vibrate about 2 400 times in a nanosecond, a long light wave of red light will vibrate 400 000 times in this interval, and a short light wave of violet light will do so 800 000 times. In 1 nanosecond, in other words, light will travel the distance of 2 400 long infrared waves, or 400 000 red light waves, or 800 000 violet light waves.

STEP 20

0.000 000 000 316 seconds ($10^{-9.5}$ s)
316 picoseconds ($10^{2.5}$ ps)

At the Step-20 time-level, we encounter the lifetimes of a number of particles more massive than the proton or neutron (such particles are sometimes called "hyperons"). One of them, the "xi-zero," has a lifetime of about 300 picoseconds, while the "lambda-zero" has one of about 250 picoseconds.

In the lifetime of the lambda-zero hyperon, light will travel a distance of 7.5 centimetres (or about the length of the little finger).

In such tiny time intervals, even light cannot go far, and one of the ways of judging lifetimes in this range is by following the track of a particle and noting the distance it covers before it breaks down. Usually, such particles are so energetic when formed that they are moving at nearly the speed of light, so that from the distance they move (a few centimetres or less) their lifetime can be estimated.

STEP 21

0.000 000 000 1 seconds (10^{-10} s)
100 picoseconds (10^{2} ps)

Of the hyperons, the "xi-minus" lifetime is 174 picoseconds, that of the "sigma-minus," 165 picoseconds, of the "omega-minus," 150 picoseconds, and of the "sigma-plus," 81 picoseconds.

In 100 picoseconds, light travels about 3 centimetres, or about the length of the end-joint of the thumb.

STEP 22

0.000 000 000 031 6 seconds ($10^{-10.5}$ s)
31.6 picoseconds ($10^{1.5}$ ps)

In 31.6 picoseconds, light travels not quite 1 centimetre, or half the width of a middle finger, while 1 gram of radium experiences 1.4 radioactive breakdowns.

STEP 23

0.000 000 000 01 seconds (10^{-11} s)
10 picoseconds (10^{1} ps)

In 10 picoseconds, light travels 3 millimetres, or just a little more than the width of one of the letters on this page.

STEP 24

0.000 000 000 003 16 seconds ($10^{-11.5}$ s)
3.16 picoseconds ($10^{0.5}$ ps)

Recently, a new, very massive lepton was discovered. If the muon is just like an electron except for having a much greater mass, the new "tau particle" or "tauon" is just like an electron except for having a *still* greater mass. The mass of the tauon is about 17 1/2 times that of a muon, or about 3 600 times that of an electron. The tauon is actually more massive than any known meson, and is twice as massive as a proton.

Because it is so massive, it is particularly unstable and has a half-life no greater than 5 picoseconds, in which time light travels 1.5 millimetres.

STEP 25

0.000 000 000 001 seconds (10^{-12} s)
1 picosecond (10^{0} ps)

In 1 picosecond (a trillionth of a second), light can only travel 0.3 millimetres, or 300 micrometres. If the movement of light in this interval of time were marked out as a line on a piece of paper, it would take a magnifying glass to distinguish it from a dot. In this interval of time, light will traverse 2.4 of the longest infrared waves, or 400 red light waves, or 800 violet light waves.

STEP 26

0.000 000 000 000 316 seconds ($10^{12.5}$ s)
316 femtoseconds ($10^{2.5}$ fs)

Cameras are devices in which flashes of light produce chemical reactions which, in turn, produce patterns of light and dark or of color that make up images of the objects that have emitted or reflected the light. Cameras have been developed that can respond to briefer and briefer flashes of light, and the briefest flash that can be currently trapped and made to yield an image is 500 femtoseconds.

In that time, light will move through a distance of 150 micrometres. Such a distance is equal to 200 times that of a red light wave, or 400 times that of a violet light wave. Of course, that is the distance that light travels *while the flash is in progress.* Once the flash is done, that length of light (a small dot of light if the eye could see it) continues to travel from the point of emission to the film within the camera.

STEP 27

0.000 000 000 000 1 seconds (10^{13} s)
100 femtoseconds (10^{2} fs)

In this length of time, light will travel a distance of 30 micrometres, about twice the width of a typical human cell; or the length of 40 red light waves, or 80 violet light waves.

STEP 28

0.000 000 000 000 031 6 seconds ($10^{-13.5}$ *s*)
31.6 femtoseconds ($10^{1.5}$ *fs*)

At this level, we are still in the range of human endeavor.

The "laser" is a device whereby energy is converted into a group of photons all of the same wavelength, and all traveling in exactly the same direction (coherent light). Lasers can be used to produce very short pulses of light, and in 1982, a laser pulse lasting only about 30 femtoseconds was produced.

In this time, light can only travel about 9 micrometres, a distance equal to half the width of a typical human cell, or to about 12 red light waves, or 24 violet light waves.

STEP 29

0.000 000 000 000 01 seconds (10^{-14} *s*)
10 femtoseconds (10^{1} *fs*)

A hyperon known as "sigma-zero" has a lifetime of about 10 femtoseconds, so that in the time that the shortest (so far) laser pulse exists, there is time for three sigma-zeroes to come into existence and break down, one after the other. In 10 femtoseconds, light travels about 3 micrometres, a little more than the width of a human sperm cell.

STEP 30

0.000 000 000 000 003 16 seconds ($10^{-14.5}$ *s*)
3.16 femtoseconds ($10^{0.5}$ *fs*)

At the Step-30 time-level, a ray of light travels only visible–light wave distances. In 3.16 femtoseconds, it travels 950 nanometres, which is the length of 1 1/4 red light waves or 2 1/2 violet light waves. However, there are electromagnetic waves shorter than those of visible light, so that in 3.16 femtoseconds, light travels the length of 95 of the shortest ultraviolet radiation.

STEP 31

0.000 000 000 000 001 seconds (10^{-15} *s*)
1 femtosecond (10^{0} *fs*)

In 1 femtosecond (a quadrillionth of a second), a ray of light travels 300 nanometres, or not more than four fifths the wavelength of the shortest-wave visible light, but 30 light waves of the shortest ultraviolet radiation.

STEP 32

0.000 000 000 000 000 316 seconds ($10^{-15.5}$ s)
316 attoseconds ($10^{2.5}$ as)

A ray of light, in this time, will travel the length of 9.5 wavelengths of the shortest ultraviolet radiation. Here again, however, there are radiations of still shorter wavelength. In this same time, light can travel the length of 950 wavelengths of the shortest-wave X rays.

STEP 33

0.000 000 000 000 000 1 seconds (10^{-16} s)
100 attoseconds (10^{2} as)

The "neutral pion" has a lifetime of about 100 attoseconds, during which time light will travel 30 nanometres, a distance equal to about the width of a nucleic acid molecule, or 300 wavelengths of the shortest-wave X rays.

It is becoming increasingly difficult to say interesting things about these shrinking time intervals. Let us, therefore, begin to skip—

. . .

STEP 37

0.000 000 000 000 000 001 seconds (10^{-18} s)
1 attosecond (10^{0} as)

In 1 attosecond (a quintillionth of a second), light will travel only 300 picometres, which is bringing us down to atomic dimensions. The light ray will manage to cross the width of three hydrogen atoms in contact, side by side, or three wavelengths of the shortest-wave X rays—the same light ray which in a second and a quarter would travel from the Earth to the Moon.

Again, there is radiation of still shorter wavelength. Even in 1 attosecond, light will travel the distance of 900 wavelengths of particularly short-wave gamma rays—

. . .

STEP 39

0.000 000 000 000 000 000 1 seconds (10^{-19} s)
0.1 attoseconds (10^{-1} as)

There are no prefixes in the SI version that signify something smaller than a quintillionth of a basic unit, so we must speak of a tenth of an attosecond, which represents the lifetime of an "eta-meson." During the lifetime of an eta-meson, light has time to move only 30 picometres, about a third of the way across a hydrogen atom, or across 90 wavelengths of short gamma rays—

. . .

STEP 47

0.000 000 000 000 000 000 000 01 seconds (10^{-23} s)
0.000 01 attoseconds (10^{-5} as)

There are particles which have been detected that have lifetimes of no more than a hundred-thousandth of an attosecond or a tenth of a trillionth of a trillionth of a second. During this lifetime, light can travel only the width of a proton.

In a sense, there is no use going further. Everything that happens is the result of subatomic particles interacting, and nothing can happen in less time than it takes light to get from one particle to another. It takes no less than 10^{-23} seconds for light to go from one particle to another even when they are in contact, so this would seem to be the shortest time in which anything can happen.

This, however, is only true of the Universe as it exists today and of the particles that now make it up. At one time—

. . .

STEP 87

0.000 000 000 000 000 000 000 000 000 000 000 000 000 000 1 seconds (10^{-43} s)
0.000 000 000 000 000 000 000 000 1 attoseconds (10^{-25} as)

In the first ten-thousandth of a second after the big bang, the Universe was too small for protons and neutrons to exist, and it consisted chiefly of quarks. As scientists probe closer and closer to the big bang (in theory), the Universe appears to have been ever smaller and hotter, and 10^{-25} attoseconds after the instant of the big bang, the Universe could have been so small and hot that even quarks could not exist.

Closer to the big bang, even the most current theories do not allow physicists to go, so at the Step-87 time-level, we must stop—at a stage where light can travel only a hundreth of a millionth of a trillionth of the width of a proton and where this may have been the full width of the Universe at the time.

In 163 steps, then, covering 81 1/2 orders of magnitude, we go from the incomprehensibly small interval of time that may have existed between the big bang and the formation of quarks, to the incomprehensibly long interval of time that may represent the half-life of the proton.

The Ladder of Speed

UPWARD

UPWARD

STEP 1

1 metre per second (10^0 m/s)
3.6 kilometres per hour ($10^{0.55}$ km/h)

In the ladder of length, I occasionally dramatized distance by indicating how long it would take for a ray of light, or some other speedy object, to traverse it. In the ladder of time, I would dramatize duration by indicating how far a ray of light, or some other speedy object, could travel in that period.

In either case, I was discussing speed, the rate of change of position with time, that is, the distance covered by a moving object in a given period of time. Since the basic unit of distance, or length, in the SI version of the metric system is the metre and the basic unit of time is the second, the unit of speed in the SI version is the "metre per second" (symbolized "m/s"), that is, the number of metres an object can move in one second.

Since there are 3 600 seconds in an hour, an object moving at a steady speed of 1 metre per second would move 3 600 metres per hour, or 3.6 kilometres per hour (symbolized "km/h"). Kilometres per hour is not really an SI unit, but its use is permitted because it is much more convenient to discuss the speed of ordinary vehicles in this way than in metres per second.

A speed of 1 metre per second, in terms familiar to Americans, is 2.24 miles per hour, and we can recognize this as the speed of a leisurely walk. At 1 metre per second, one can walk a city block in 1 1/3 minutes; yet when I walk the streets of Manhattan, experience has taught me that I need not allow more than 1 minute per city block (if we don't count the delay of red lights) when I am walking at my normal clip.

My own walking speed, therefore, is about 1.33 metres per second, or 4.68 kilometres per hour (which is equivalent, in American units, to just about 3 miles per hour).

When in a rush, I can walk at perhaps 6 1/2 kilometres per hour, but I find it possible to keep that up for only a few blocks, whereas I can maintain the lower speed without undue trouble for a couple of kilometres.

Ordinarily, we move through air, which offers very little resistance to our progress. When we are swimming, we are pushing our way through water, which is a much more resistant medium.

However quickly you may seem to be swimming, to yourself and to those watching and cheering you on, you are progressing at no better than a walk, and (usually) a slow walk at that. Indeed, the fastest swimming speed recorded, as part of a record 100-metre race, is 7.92 kilometres per hour, or 2.2 metres per second.

The world record for a 1.5-kilometre swim is 14 minutes 58.27 seconds, which represents an average speed of 6.01 kilometres per hour, or 1.67 metres per second. Undoubtedly, I could not keep up with a champion swimmer for long, if I were walking alongside the pool, but a younger man in better shape could undoubtedly do so.

STEP 2

3.16 metres per second ($10^{0.5}$ m/s)
11.4 kilometres per hour ($10^{1.05}$ km/h)

In walking, as in every human activity, there are specialists who can far outdo the average person. There are people who can walk very quickly over long distances (it remains a walk, whatever the speed, if one foot is always touching the ground).

The official world record for walking a distance of 50 kilometres is 3 hours 41 minutes and 39 seconds. This represents an average speed of 13.535 kilometres per hour, or 3.76 metres per second. This is about 2 1/2 times as fast as the average walking speed.

Naturally, one can go faster over a shorter distance than a longer, and the official record for a 20-kilometre walk is 1 hour 20 minutes and 6.8 seconds. The average speed here is 14.981 kilometres per hour, or 4.16 metres per second. This is 3 times as fast as the average walking speed, and nearly 2 times as fast as the fastest human swimming speed.

STEP 3

10 metres per second (10^1 m/s)
36 kilometres per hour ($10^{1.55}$ km/h)

When we move up to the Step-3 speed-level, we must leave walking and swimming behind, where human beings are concerned, and deal with running. Running differs from walking in that, when running, both feet may be off the ground at given moments. On the whole, running is an easier and less energy-consuming means of progression at higher speeds than walking is.

The most popular of the long races is the marathon, which is 42.19 kilometres long (26 miles 385 yards is the standard length in American units—supposedly the distance from Marathon to Athens, which was run, in 490 B.C., by Pheidippides, who gasped out the news of the Athenian victory over the Persians, and died).

How long it took Pheidippides to make that run, we don't know, but the modern record for a marathon run is 2 hours 8 minutes 33.6 seconds. This represents an average speed of 19.69 kilometres per hour (5.47 metres per second), and is probably 1 1/3 times as fast as a champion walker could manage over the same distance.

Shorter distances are run at a greater average speed. Thirty years ago, the 4-minute mile seemed an unrealizable dream, but now top-notch racers routinely do a mile in less than 4 minutes. The world record is 3 minutes 48.8 seconds, for an average speed of 25.32 kilometres per hour, or 7.03 metres per second.

The world record for racing 100 metres is 9.95 seconds, and the average speed over that short distance is 36.182 kilometres per hour (10.05 metres per second), which is almost exactly at the Step-3 speed-level. Such a race includes a few metres at the start, in which the racer accelerates, and a few metres toward the end, in which the racer has tired. The middle 50 metres would see him racing at a clip of perhaps 43 kilometres per hour, or 11.94 metres per second.

This may seem like the ultimate speed for human beings by muscle power alone, but parts of the human body can move faster than the entire body can. At least the speed of the human arm in a rapid karate chop has been reported as up to 51.8 kilometres per hour (14.4 metres per second).

Human beings can also increase their speed if they cut down the friction of the surface and glide rather than run. A human being on roller skates has been timed up to 41.5 kilometres per hour (11.5 metres per second)—no faster than a racer can go, but over longer distances, surely. On ice skates, the fastest racing speed (over a course of 500 metres) is 48.78 kilometres per hour (13.55 metres per second).

In judging how quickly nonhuman organisms can move, we tend to overestimate the speed of small organisms and underestimate the speed of large ones. We may be judging, perhaps, by the time it takes to move a distance equal to the length of the animal's body. A chipmunk seems to scurry and an elephant to lumber, but an elephant can run far faster than a chipmunk. An African elephant is not usually thought of as a fast animal, but it has a long stride, and it is reported to be capable of achieving a speed of 39.5 kilometres per hour (11 metres per second), and this is almost as fast as a human being can run under record conditions.

We also tend to overestimate speed where an unfamiliar method of locomotion is used. Because a snake lacks legs we are astonished at its crawling progress, and, perhaps judging how slowly we would go at a wriggle, we assume a snake is moving at a far greater speed than it is. Actually, the fastest speed for a snake on a level surface would be that of a black mamba, which may reach a speed of 24 kilometres per hour (6.7 metres per second). The fastest land reptile of any kind may be a kind of lizard known as a race runner, which can reach 29 kilometres per hour (8 metres per second).

Oddly enough, some reptiles can do better in water. There are leatherback turtles that can reach speeds of 35.4 kilometres per hour under water (9.8 metres per second), which would make them the fastest reptiles of any kind. (This is surprising since their land-based relatives—the tortoises—are notorious for their slowness.)

The fastest avian swimmers are the penguins, some of which can swim up to 36 kilometres per hour (10 metres per second). The fastest pinniped, the sea lion, can swim up to 40 kilometres per hour (11.1 metres per second).

STEP 4

31.6 metres per second ($10^{1.5}$ m/s)
114 kilometres per hour ($10^{2.05}$ km/h)

At the Step-4 speed-level, we leave unaided human speeds far behind, but we have not totally outraced other mammals.

The animals we most familiarly equate with speed beyond the

human is the thoroughbred racehorse, on whose abilities millions of human beings bet millions of dollars daily.

The world record for a horse running 3 miles is 5 minutes 15.0 seconds, which is the equivalent of 55.17 kilometres per hour (15.33 metres per second). Over the space of a quarter of a mile, however, for which the horse-racing record is 20.8 seconds, the speed represented is 69.61 kilometres per hour (19.33 metres per second). In other words, the best a horse can do over a short distance is about 1 5/8 as well as the best a human being can do. (Most people would probably be surprised that a horse surpasses a human being by so little in this respect. It is not that they don't realize how fast a horse can run, but that they don't realize how fast a man can run.)

In the case of horses, we are, of course, talking about animals bred for speed, exercised, and exquisitely cared for. Their accomplishments in this respect are far beyond the average equine capacity. (This is also true of greyhounds, which have reached speeds of 67.1 kilometres per hour, or 18.65 metres per second, just under the best racehorse speed, and without the advantage of the much larger equine stride.)

What about animals in the wild, which are not specially selected, or bred, or trained by human hands, and which are timed in haphazard fashion so that the fastest may just happen to escape our notice?

A California jackrabbit can reach a speed of 65 kilometres per hour (nearly 18 metres per second). This is very close to the best a racehorse can do. A killer whale may achieve this very speed, but will do so through the more resistant medium of water (though its streamlining virtually cancels that resistance and its flukes are beautifully adapted to propulsion).

The ostrich can possibly reach a speed of 72 kilometres per hour (20 metres per second), thus outdoing by a bit the best that a horse can do.

None of these, however, can get very close to the Step-4 speed-level. To approach the level, we must pass on to the antelopes. The champion of the group is the pronghorn antelope, which can manage a speed, according to reports, of very close to 100 kilometres per hour (27.5 metres per second).

Just the same, in short bursts, even the pronghorn antelope can be outdone by the cheetah, which is the fastest land animal alive, and for which speeds of up to 101 kilometres per hour (28 metres per second) have been claimed.

Yet in water, the cheetah can be outdone by some fish. Various swordfishes seem to be the fastest fish and, indeed, the fastest nonflying creatures of any kind. One variety, the sailfish, has been reported to move at a speed of as much as 109.7 kilometres per hour (30 metres per second). This speed is 1 2/3 times the best mark of a killer whale.

The sailfish is, in fact, faster than most flyers.

The fastest insect, the fastest bat, even the fastest flying fish (while gliding through the air) all achieve a maximum speed over short distances in the neighborhood of 55 kilometres per hour (15 metres per second), only half the speed of a sailfish.

And yet even a sailfish does not quite reach the Step-4 speed-level, and it must take a back seat in this respect to certain birds.

The canvasback duck, in level flight, can reach speeds as high as 116 kilometres per hour (32 metres per second), just topping the Step-4 speed-level.

The fastest of all birds is the aptly named spine-tailed swift, which has been clocked at a speed of 171 kilometres per hour (47.5 metres per second). The spine-tailed swift can move 1.7 times as fast as a cheetah; 2.4 times as fast as the best a racehorse can do; and about 4 times as fast as the best a human being can do.

For human beings to match the sailfish at sea is not easy, even with all of technology at their disposal. The fastest yacht moves at a speed of 66.8 kilometres per hour (18.5 metres per second); the fastest passenger liner, at 71.0 kilometres per hour (19.7 metres per second); the fastest destroyer, at 83.4 kilometres per hour (23.2 metres per second). The fastest warship of any kind is a hydrofoil that has managed to attain a speed of 113 kilometres per hour (31.4 metres per second).

The fastest motorboat can travel 156.4 kilometres per hour (43.4 metres per second). This finally well outdoes the sailfish and, in fact, gets close to the spine-tailed swift in the air.

On land, human beings can do pretty well, thanks to wheels, ice, and the pull of gravity, even without the aid of motors.

Thus, the fastest speed on a bicycle is 94.35 kilometres per hour (26.2 metres per second) if one person is peddling, and 101.24 kilometres per hour (28.1 metres per second) if more than one person is. On a toboggan, speeds of 101.5 kilometres per hour (28.2 metres per second) have been obtained. Cyclists and tobogganists, in other words, can (at their best) match a cheetah.

But human beings can do distinctly better than the cheetah, too, even without motors. A glider, over a 100-kilometre course, has averaged a speed of 165.3 kilometres per hour (45.9 metres per second). This nearly matches the spine-tailed swift.

STEP 5

100 metres per second (10^2 m/s)
360 kilometres per hour ($10^{2.55}$ km/h)

Human beings, without motors, can outdo the swift, too. The fastest speed claimed for a skier (zooming downhill) is 200.2 kilometres per hour (55.6 metres per second). An iceboat, however, on the level and driven by the wind, is reported to have reached the speed of 230 kilometres per hour (64 metres per second). This is 1 1/3 times the speed of a spine-tailed swift, and is probably the fastest that human beings can move without motorized transport and under controlled conditions that leave them expectations of survival.

Human beings can make inanimate objects travel as fast as a swift, or faster, by the use of muscle power alone.

A baseball pitcher can throw a baseball as fast as 162.35 kilometres per hour (45.1 metres per second); a hockey puck has been found to move as fast as 190 kilometres per hour (53 metres per second); a tennis ball as fast as 263 kilometres per hour (73 metres per second); and a golfball as fast as 273.5 kilometres per hour (765 metres per second). These last two outdo an iceboat in speed.

The fastest object in any recognized sport, however, is the ball in the game of jai alai. Such a ball has been reported as moving at speeds as high as 302.5 kilometres per hour (84 metres per second). Such a ball moves 1.77 times as fast as a spine-tailed swift, and 4 1/3 times as fast as the best a racehorse can do.

—And yet there are living organisms that can do better still.

In considering the speeds of flying birds, we have been dealing with flights on the level. Predatory birds, however, sometimes dive after their prey, and with the pull of gravity adding to their speed, they might outdo a jai alai ball at its fastest.

A diving peregrine falcon is reported to have been timed at 350 kilometres per hour (97 metres per second), and there is some speculation that a diving speed of up to 385 kilometres per hour (107 metres per second) could be attained. If so, that would seem to be the fastest any living thing could travel, in controlled fashion, without motorized transport, and it is just at the Step-5 speed-level.

We must move on to inanimate objects, however, and the one inanimate object that impressed even primitive man with its speed is, of course, the wind. "Fast as the wind" is a common cliché.

The speed of the wind is particularly noticeable in storms (a hurricane is defined, for instance, by an associated wind that is above 120 kilometres per hour, or 33.5 metres per second). The strongest wind gust recorded during a hurricane reached a speed of 317 kilometres per hour (88 metres per second).

The fastest winds are those whirling about in a tornado. There, wind speeds of 480 kilometres per hour (134 metres per second) have been recorded, and the speed of a diving falcon is exceeded.

Winds tend to be more rapid at higher elevations, where the air is less dense and resistant, and where a smaller gas-mass need be moved. At the top of Mount Washington, winds of 372 kilometres per hour (103 metres per second) have been recorded. In the jet stream (a fast stream of air circling the Earth at a height of about 45 kilometres above the surface), a wind speed of 656 kilometres per hour (182 metres per second) has been observed, and this is faster than the wind in a tornado.

"As fast as the wind" can be applied, even at tornado levels, to human motorized transport by land. Passenger automobiles can attain speeds of up to 310 kilometres per hour (87 metres per second). There are trains, moving along conventional rails in France, that can reach speeds of 380 kilometres per hour (105 metres per second). In Japan, experiments are being conducted with monorails, where the carriages are magnetically levitated so that they make no solid-solid contact and friction recedes to very low levels. In that case, speeds of up to 515 kilometres per hour (140 metres per second) are attained, and the passengers would outspeed a tornado wind.

In the air, the fastest helicopter can register a speed of 368 kilometres per hour (102 metres per second).

STEP 6

316 metres per second ($10^{2.5}$ m/s)
1 140 kilometres per hour ($10^{3.05}$ km/h)

Before World War II, the fastest means of human transport were piston-engined automobiles on land and propellor-driven airplanes in the air.

With modern improvements, piston-engined automobiles have now reached extreme speeds of 673 kilometres per hour (187 metres per second).

The record for propeller-driven airplanes is 878 kilometres per hour (244 metres per second). This is 2 1/4 times as fast as a diving falcon, so that with such airplanes, human beings outflew the birds and became the fastest living things.

Even the fastest propeller-driven planes, however, have speeds only three fourths that of the Step-6 speed-level. To reach that level, we must turn to the jets and rockets developed during and soon after World War II. For instance, rocket-powered land vehicles have gone as fast as 1 190 kilometres per hour (330 metres per second), and that is just beyond the Step-6 speed-level.

The speed of sound through the air under Earth-surface conditions is 1 195 kilometres per hour (332 metres per second). The rocket-powered land vehicle just referred to goes at almost exactly the speed of sound and therefore moves at "Mach 1" (named for Ernst Mach, an Austrian physicist).

At this level, we can also get astronomical. This is not to deny that some astronomical speeds are quite ordinary. For instance, Venus rotates on its axis so slowly, relative to the stars, that a point on its equator moves at a speed of barely 6.5 kilometres per hour—no more than the speed of a fast walk. Places elsewhere on the surface move even more slowly.

We are interested, however, in astronomic speeds that are fast compared to the familiar speeds of Earth, or even of most of human technology. Thus a spot on Mars's equator moves at a speed of 870 kilometres per hour (241 metres per second), which is about as fast as the fastest propeller airplane can go.

Earth rotates just a trifle faster than Mars does, but Earth is considerably larger, so that a spot on Earth's equator must travel considerably faster than one on Mars's equator, to make its way around the larger circle in the same time. Earth's equatorial speed is 1 680 kilometres per hour (467 metres per second), which is well over the Step-6 level.

Earth's equatorial speed is faster than sound, and, indeed, represents a speed of Mach 1.4. We know that to go faster than sound creates a sonic boom, and there may be some surprise that Earth's rotation does not. However, the sonic boom is created when the speeds faster than sound are speeds relative to the air. The Earth's rotation *carries the air along with it,* so that Earth's equatorial speed relative to the air is zero (except for what wind may exist).

STEP 7

1 000 metres per second (10^3 m/s)
1 kilometre per second (10^0 km/s)

At this stage, I abandon kilometres per hour as a unit. Kilometres per second is far more dramatic, and is, moreover, strictly proper from the standpoint of the SI version of the metric system.

Thus, the record speed attained by a jet plane (set in 1976) is

0.9802 kilometres per second, almost three times the speed of sound (Mach 3) and very nearly at the Step-7 speed-level. The fastest rocket-powered object on rails has attained a speed of 1.38 kilometres per second (Mach 4.15).

The orbital speed of the Moon about the Earth is, on the average, 1.02 kilometres per second. That of Deimos about Mars is 1.17 kilometres per second. Deimos is considerably closer to Mars than the Moon is to Earth, and should for that reason (all things being equal) experience a stronger gravitational pull. All things are not equal, however. Mars is a smaller world than Earth is and has a smaller gravitational field. Deimos's nearness to its planet is cancelled by the planet's small size.

STEP 8

3 160 metres per second ($10^{3.5}$ m/s)
3.16 kilometres per second ($10^{0.5}$ km/s)

We are now beyond the speed-levels of all human vehicles except rocket ships, and we will have to confine ourselves to those, and to astronomic phenomena.

The orbital speed of Phobos, the innermost of Mars's two satellites, is 2.13 kilometres per second.

Phobos is very close to Mars, while the planet Pluto is the farthest from the Sun of all known planets. The gravitational field of the Sun is so enormous compared to that of Mars, however, that Pluto has a faster orbital speed relative to the Sun than Phobos has relative to Mars. The average orbital speed of Pluto is 4.7 kilometres per second.

The escape velocity is the speed an object must have to move indefinitely far from a world despite the latter's gravitational pull, even without the input of additional power. It is also the maximum speed attained by an object falling to a world's surface from a position of rest.

Thus, the escape velocity from the Moon is 2.38 kilometres per second; while from Ganymede, the largest known satellite, it is 2.75 kilometres per second; and from Mercury, a planet smaller than Ganymede but more massive, it is 4.2 kilometres per second.

STEP 9

10 000 metres per second (10^4 m/s)
10 kilometres per second (10^1 km/s)

The escape velocity from Earth is 11.3 kilometres per second. The minimum speed required for maintaining an orbit about the Earth just beyond its atmosphere is its escape velocity divided by the square root of 2 (which is 1.414 . . .). The minimum orbital speed of an object in near orbit about the Earth is therefore 8 kilometres per second. Thus, Yuri Gagarin, the first man to orbit the Earth in a rocket ship, attained a maximum speed of 8.05 kilometres per second, and traveled 8.2 times as fast as the fastest jet plane. This is still only four fifths of the way to the Step-9 speed-level.

Greater speeds were involved in the vehicles that carried men to the Moon, where 11.1 kilometres per second was attained. This is just beyond the Step-9 speed-level, and represents the greatest speed at which (as yet) human beings have traveled. This is roughly 100 times as fast as any nonhuman living thing has ever traveled under its own power.

Orbital speeds go up, all things being equal, as the mass of the central body increases. Thus the satellites of Saturn and Jupiter are quite likely to move more quickly than the satellites of Mars and Earth. Dione moves about Saturn at an orbital velocity of 10.0 kilometres per second, while Ganymede moves about Jupiter at 10.9 kilometres per second.

Saturn itself moves about the Sun at an average orbital velocity of about 9.6 kilometres per second.

Saturn and Jupiter, although the largest planets, have the shortest periods of rotation. The equatorial velocities of these two planets are therefore much faster than Earth's—10.3 kilometres per second for Saturn and 12.5 kilometres per second for Jupiter. Jupiter's is the fastest equatorial velocity of any known planet.

STEP 10

31 600 metres per second ($10^{4.5}$ m/s)
31.6 kilometres per second ($10^{1.5}$ km/s)

We are now at a speed-level approaching the limit of the orbital velocities of satellites. The innermost particles of those portions of Saturn's rings visible from Earth circle Saturn at velocities of 22.2 kilometres per second. Amalthea, the innermost satellite of Jupiter (of those visible from Earth), has an orbital velocity of 27.8 kilometres per second. If we were to consider an object orbiting Saturn just outside its atmosphere, it would have an orbital speed of 25 kilometres per second. An object orbiting Jupiter just outside its atmosphere would have an orbital speed of 43 kilometres per second, so that it is the only planet that might conceivably have a satellite with an orbital velocity beyond the Step-10 speed-level.

Where planetary orbital velocities are concerned, that of Earth about the Sun is 29.8 kilometres per second, and that of Venus is 35.1 kilometres per second. There we are approaching an upper level, too.

We are doing the same with planetary escape velocities. The escape velocity from Saturn (from the visible level of its cloud layer) is 35.2 kilometres per second.

STEP 11

100 000 metres per second (10^5 m/s)
100 kilometres per second (10^2 km/s)

At this level, we are beyond planetary speeds of any kind. Even Mercury at its closest to the Sun is moving at an orbital speed of only 56 kilometres per second. Icarus, a small asteroid that approaches the Sun more closely than Mercury, may reach a speed of 70 kilometres per second. There are some comets that approach the Sun still more closely than Icarus does, and the closest, one which skimmed the Sun's surface in 1963, may have reached a speed of 100 kilometres per second.

As for planetary escape velocities, the greatest is that of Jupiter, from the level of its visible cloud-top surface, and that is only 60.5 kilometres per second.

The fastest rocket human beings have launched is Helios B, and that attained a speed as high as 66.65 kilometres per second, relative to Earth.

We can move on to stellar speeds. Lacaille 9352, a star in our own neighborhood of the Galaxy, moves at a speed of 119 kilometres per second relative to the Sun.

STEP 12

316 000 metres per second ($10^{5.5}$ m/s)
316 kilometres per second ($10^{2.5}$ km/s)

The fastest moving nearby star (relative to the Sun) is Kapteyn's star, which is 13 light-years away from us. It is moving at a speed of 294 kilometres per second.

Both the Sun and Kapteyn's star (and all the other stars in the Sun's neighborhood, and in the outskirts of the Galaxy generally) are moving in orbits about the center of the Galaxy. The orbits are eccentric, and are inclined to the plane of the Galaxy in varying degrees. It is these orbital differences that make the individual stars show motion relative to each other, even as all together move about the Galactic center. The Sun is moving around the Galactic center in a fairly circular orbit, and at a speed of about 220 kilometres per second.

STEP 13

1 000 000 metres per second (10^6 m/s)
1 megametre per second (10^0 Mm/s)

The only speed connected with the Solar system that approaches the Step-13 level is the escape velocity from the Sun's visible surface. This is 0.617 megametres per second.

Ordinary main-sequence stars more massive than the Sun could have a larger escape velocity but not much larger. The additional mass increases the escape velocity, but such massive stars are also more voluminous, and the greater distance of the surface of the star from the center tends to decrease the escape velocity. Some stars may have escape velocities as great as 0.825 megametres per second, but it is doubtful if any of them reach the Step-13 speed-level.

We can, however, reach that level by returning to rotational velocities, though not in the Solar system. A point on the equator of the neutron star within the Crab Nebula, the one with the fastest known rotation, moves at a speed of 1.33 megametres per second.

STEP 14

3 160 000 metres per second ($10^{6.5}$ m/s)
3.16 megametres per second ($10^{0.5}$ Mm/s)

The Universe is expanding. This means that the distant galaxies are all moving away from us. What is more, since the Universe as a whole is expanding, the speeds of these recessions, relative to ourselves, is cumulative. The farther away a galaxy is from us, the faster it appears to be receding.

It is currently estimated that for every increase of one million parsecs in distance, there is an increase of 50 kilometres per second in speed of recession. That means that a galaxy that is 63 million parsecs away from us (90 times as far away as the relatively nearby Andromeda galaxy) would be receding from us at a speed of 3.16 megametres per second.

STEP 15

10 000 000 metres per second (10^7 *m/s*)
10 megametres per second (10^1 *Mm/s*)

Although main-sequence stars do not have escape velocities in the megametre per second range, collapsed stars can. A white dwarf can have a mass as large as our Sun does and yet have a diameter no greater than that of a small planet. The gravitational field is much more intense at the surface of a white dwarf, therefore, than at the surface of a star like our Sun, since the white dwarf's surface is so much closer to its center than the Sun's is. Thus, Sirius B, with a mass equal to 1.05 times that of the Sun and a diameter of 11.1 megametres (less than that of the Earth, and only 1/125 that of the Sun) has an escape velocity of about 7 megametres per second.

A galaxy 200 million parsecs from us would be receding at about 10 megametres per second. Such a galaxy would be so far away from us that it would only be visible in the best of modern light-telescopes.

Closer to home—when radioactivity was discovered in the 1890s, substances such as uranium and thorium were found to emit radiations. Some of this radiation, called "alpha rays," turned out to consist of very energetic "alpha particles," which, in turn, were found to be the nuclei of helium atoms.

So energetic were these particles that they moved at enormous speeds—beyond any that had been observed up to that point. Even a relatively slow alpha particle, as it emerged from a radioactive substance, would have a speed of about 14 megametres per second.

STEP 16

31 600 000 metres per second ($10^{7.5}$ *m/s*)
31.6 megametres per second ($10^{1.5}$ *Mm/s*)

Some particularly energetic alpha particles speed at up to 22 megametres per second. Distant objects 600 million parsecs from us recede at a speed of 30 megametres per second, or so. Ordinary galaxies are not visible at such distances; only quasars are.

STEP 17

100 000 000 metres per second (10^8 *m/s*)
100 megametres per second (10^2 *Mm/s*)

Distant quasars, lying more than a billion parsecs from us, recede at a speed of 100 megametres per second or more.

Some radioactive substances give off beta particles (speeding electrons) that are 1/7 300 as massive as alpha particles and can therefore be accelerated to much higher speeds by the energy available in radioactive breakdowns. Even relatively slow beta particles travel at about 160 megametres per second.

The escape velocity from a neutron star, which packs all the mass of an ordinary star like the Sun into a tiny globe perhaps 14 kilometres across, is about 200 megametres per second.

Light travels at a speed that depends on the refractive index of the material it travels through. Diamond has a comparatively high index of refraction (which accounts for the way it sparkles and shows colors when properly faceted and held at varying angles). Light traveling through diamond moves at a speed of 124 megametres per second.

STEP 18

316 000 000 metres per second ($10^{8.5}$ m/s)
316 megametres per second ($10^{2.5}$ Mm/s)

Light travels at its maximum speed when racing through a vacuum. Its speed is then just short of 300 megametres per second, or nearly the Step-18 distance.

It is now well established that nothing we have ever observed can go faster than the speed of light in a vacuum, so that a true Step-18 speed would seem to be forever unattainable. (There have been speculations that there are particles—named "tachyons," from a Greek word for "fast"—that can go faster than light at any speed up to the infinite. Such particles, however, have not yet been detected, and few scientists are persuaded that they exist.)

The farthest known quasars may be receding from us at speeds of 270 megametres per second or so; and the most energetic subatomic particles we know of may be speeding at 290 megametres per second, or so—but no matter how such objects may approach the speed of light, they never quite reach it, let alone surpass it.

Any particles other than photons (the fundamental particles of light and of other electromagnetic radiation) which, like photons, have a rest mass of zero—such as gravitons and neutrinos—would also travel at the speed of light.

And so, in 17 steps, covering eight-and-a-half orders of magnitude, we have gone from the speed of a slow human walk to the ultimate speed of light in a vacuum.

The Ladder of Speed

DOWNWARD

RJONES

DOWNWARD

STEP 1

1 metre per second (10^0 m/s)

We are now back to 1 metre per second, the speed of a human stroll.

Astronomical speeds (relative to some convenient reference point) are, for the most part, faster than this, but there are some that are as slow, or slower.

Thus, the most slowly rotating planet is Venus. It rotates on its axis, relative to the stars, once in 243.09 Earth-days. That means that a point on its equator moves at only 1.82 metres per second. This is the rate of a brisk walk.

If you were standing on the surface of Venus at its equator, and could somehow endure its horrible surface conditions, and could see through its clouds, you would observe the stars taking 121.54 days to march across the sky from rising to setting. If you walked briskly in the direction opposite to that in which Venus was rotating, you would cancel out the stellar motion. A star that was overhead would therefore remain overhead for as long as you continued to walk briskly. (In saying this, I am ignoring the effect of Venus's revolution about the Sun.)

That, however, is the rotational speed at Venus's equator. As one goes north or south from the equator, the surface makes smaller circles in rotating, but takes the same time to do so, and therefore moves more slowly. At a latitude of 57.5° north (or south), the rotational speed on Venus would be just about 1 metre per second. Farther north (or south), it would be slower still.

The same is true on any rotating body. On Earth, which rotates much more rapidly than Venus, a spot that is 12 kilometres from the North Pole (or the South Pole) rotates about Earth's axis at a speed of only 1 metre per second.

STEP 2

0.316 metres per second ($10^{-0.5}$ m/s)
3.16 decimetres per second ($10^{0.5}$ dm/s)

Just as there are animals that can move more quickly than human beings, there are others which can only move more slowly—even when they appear to move quickly.

A centipede, for instance, can appear to be moving very quickly indeed as it scurries along, with all its legs pumping, and covering a distance equal to its own body length in little time (to say nothing of its progress being hidden in the underbrush, so that it seems very soon lost to sight). Nevertheless, its motion has been timed at no more than 5 decimetres per second, less than a third the speed of a brisk human walk.

STEP 3

0.1 metres per second (10^{-1} m/s)
1 decimetre per second (10^{0} dm/s)

Proverbial for slowness are the tortoises (which is the reason for its choice as one of the contestants in the famous fable of the tortoise and the hare). The giant tortoise of the Galapagos Islands apparently cannot move faster than 0.75 decimetres per second. In an hour, in other words, such a tortoise can move forward about 3 1/3 city blocks, if it maintains this breakneck speed in a straight line during all that time.

A three-toed sloth, the slowest of mammals (though not well named, for it moves slowly not out of laziness but out of inability to move faster) progresses along the branches of trees (while suspended upside down) at about this same speed at best.

STEP 4

0.031 6 metres per second ($10^{-1.5}$ m/s)
3.16 centimetres per second ($10^{0.5}$ cm/s)

The sloth in the trees is in its element and, slow as it is, that is when it is moving most quickly. A sloth on the ground is handicapped and is slower still. It can then move, according to reports, at no better than 4.5 centimetres per second—or 2 city blocks per hour.

STEP 5

0.01 metres per second (10^{-2} m/s)
1 centimetre per second (10^{0} cm/s)

Another creature well known for its slowness is the garden snail, which glides slowly forward on its "foot" at a speed no greater than 1.35 centimetres per second at the most. It would take such a racing snail two hours to move a city block.

Living organisms capable of motion can usually move, voluntarily, at any speed less than their maximum, right down to motionlessness. It would be uninteresting to cite them in the descent of the ladder of speed. Let us, rather, switch to a few kinds of inanimate motions that involve familiar phenomena, skipping steps as necessary.

. . .

STEP 7

0.001 metres per second (10^{-3} m/s)
1 millimetre per second (10^{0} mm/s)

On most shorelines of the world, the water rises and falls twice a day. We call these water-level changes "tides," and their existence depends on the fact that the Moon (and to a lesser degree, the Sun) exerts a greater gravitational pull on the side of the Earth facing it, than on the side that is away from it. The Earth stretches, therefore, in line with the Moon, and the ocean water bulges on either side. As the Earth turns, each spot on the surface passes through the bulge each half-day.

The extent to which the water rises and falls differs from spot to spot, depending on the shape of the coastline, the relative positions of the Moon and Sun, and so on. The greatest tides are in the Bay of Fundy, which lies between the Canadian provinces of Nova Scotia and New Brunswick. The bay is funnel-shaped so that, as the tide moves in, it is compressed into steadily narrower bounds and tends to pile up.

The extreme difference between high tide and low tide in the Bay of Fundy is 16.3 metres. Water moves from low tide to high tide (or vice versa) in about 6 hours 12 minutes. That means that the average speed of rise or fall in water level in the Bay of Fundy is about 0.73 millimetres per second. Midway between low tide and high tide, the water level is changing most quickly, and it may then reach the mark of 1 millimetre per second.

This would represent the fastest *vertical* tidal movement on Earth. Naturally, as the water level rises, the water moves inland up the sloping shore at a much greater speed, that speed depending on the actual slope of the shore.

STEP 8

0.000 316 metres per second ($10^{-3.5}$ m/s)
316 micrometres per second ($10^{2.5}$ μm/s)

Glaciers are great rivers of ice which, very slowly, move down the sides of mountains. Ice deforms under the pressure of its own great weight when it is present in huge bulk, and will spread outward and downward as a result. The speed of this spreading depends on the degree of slope, the temperature, and so on, but it is always very slow.

A glacier in Greenland is possibly the fastest known, and its movement is sometimes as fast as 280 micrometres per second. That means that in a day of this fastest movement, steadily continuing, the glacier will creep forward about a third of a city block.

. . .

STEP 18

0.000 000 003 16 metres per second ($10^{-8.5}$ m/s)
3.16 nanometres per second ($10^{0.5}$ nm/s)

As we all know, hair grows continuously. It doesn't grow quickly, but a 24-hour facial stubble on a male human being has grown long enough to be unsightly if it is not lopped off; and in five or six hours it has grown long enough to be unpleasantly rough to the touch.

The fastest-growing hair on the human body can grow at a speed of up to 4.3 nanometres per second. That is about three eighths of a millimetre in a day. In 7 3/8 years, a strand of hair (if it does not fall out in the meanwhile, as it almost surely will) would grow to be 1 metre long, at this rate.

STEP 19

0.000 000 001 metres per second (10^{-9} m/s)
1 nanometre per second (10^0 nm/s)

The Earth's crust is broken up into half a dozen huge plates, plus a number of smaller ones, that very slowly move relative to each other. These moving plates break landmasses apart, or drive them into one another. As landmasses move apart, a widening ocean may appear between them; and as they move together, a mountain range may form at the collision boundary.

Such movements are very slow indeed. Thus, South America and Africa are moving apart, and the Atlantic Ocean is widening, at an estimated speed of 1.25 nanometres per second. In one year, at that rate, the Atlantic Ocean would widen 4 centimetres.

Then, too, because of friction, the tides must expend energy in moving up and down the shore lines. This tidal friction is overcome at the expense of the rotational energy of the Earth, so that the Earth's rate of rotation is slowing in an extremely leisurely fashion. This, in turn, means that the Earth is losing some of its angular momentum.

Angular momentum cannot be truly lost; it can only be transferred. In this case, it is transferred to the angular momentum of revolution as Earth and Moon turn about a common center of gravity. For the angular momentum of revolution to increase, as the angular momentum of Earth's rotation decreases, the Moon must move farther and farther away from the Earth.

It is estimated that the rate at which the Moon is receding from Earth is just about 1 nanometre per second. This means that in the space of 1 year, the Moon recedes from Earth by 3.15 centimetres.

—And at this point we will stop our descent, for nothing of interest lies below.

The Ladder of Temperature

UPWARD

STEP 1

1 degree Celsius (10^{0} °C)
274.15 kelvin ($10^{2.438}$ K)

The most common measurement we make that involves neither length, mass, nor time, is that of temperature.

Temperature is the measure of the average energy of motion possessed by each particle of a mass. Another way of putting it is that it is the measure of the intensity of the heat of a mass.

A particular object may have a great deal of heat at a low temperature or a relatively small amount of heat at a high temperature; thus, you can have a nailhead that is red hot, and a ton of sand, with a much greater amount of heat, altogether, which is merely warm. This is analogous to the manner in which you can have a small pool of water which may be fifteen metres deep, and a large lake with a much greater amount of water altogether, that is nowhere deeper than five metres.

Energy, in the form of heat, flows from a body of higher temperature to one of lower temperature, regardless of the total heat present in either body. If you touch a red-hot nailhead, heat will flow from it to the part of your body that touches it, even though your body contains more heat, altogether, than the nailhead does. There will be a quick rise in the temperature of the part of your body in contact with the pinhead, and you will feel pain.

Temperature in everyday affairs is measured by a thermometer, which relies on the steady expansion of a thin column of mercury with rising temperature—or its contraction with falling temperature.

A Swedish physicist, Anders Celsius, devised a scale for expressing temperature. The height of the mercury column at the temperature at which water froze to ice (or ice melted to water) under ordinary conditions, he marked and set equal to 0. The temperature at which water boiled to steam (or steam condensed to water) under ordinary conditions, he set equal to 100. The temperatures between were marked off into one hundred equal divisions of "degrees" (from Latin words meaning "to step down"—that is, from 100 to 0).

Originally, this scale was referred to as the "centigrade scale" (from a Latin word meaning "a hundred steps"), so that a temperature was said to be a particular number of "degrees Centigrade." This was symbolized as °C. A couple of decades ago, the name of the scale was changed to the "Celsius scale" in honor of the inventor, so that we now speak of "degrees Celsius." The symbol remained unchanged.

The Celsius scale is now used throughout the world, except (as you might, perhaps, expect) in the United States, where an older scale devised by a German-Dutch physicist, Gabriel Daniel Fahrenheit, is used. In this "Fahrenheit scale," the freezing point of water is set at 32, and the boiling point at 212, so that between the two are 180 degrees. Temperatures on the Fahrenheit scale are measured as so many "degrees Fahrenheit," and this is symbolized as °F.

It is not difficult to convert a temperature measurement from degrees Celsius to degrees Fahrenheit, or vice versa, and I will give Fahrenheit equivalents where that seems advisable. On the whole, though, we will cling to degrees Celsius—at least for ordinary temperatures.

One can easily imagine the temperature dropping lower than the freezing point of water, in which case we must use negative values, such as "ten degrees below zero Celsius," or −10 °C. On the Fahrenheit scale, one can go below the temperature of freezing water (32 degrees Fahrenheit) and still have positive values. One must go to −17.78 degrees Celsius before reaching 0 degrees Fahrenheit, but at lower temperatures still, the Fahreneheit scale must also make use of negative numbers.

If the temperature continues to go down, this implies that the average energy content per particle also goes down, and you might suppose that if the temperature goes down far enough, the energy content drops to zero. If you suppose that, you are right. Nor does one have to go down very far. The temperature at which no further energy can be withdrawn from matter under any circumstances is −273.15 degrees Celsius (or −459.67 degrees Fahrenheit). This temperature is called "absolute zero."

William Thomson, Lord Kelvin, first suggested a temperature scale that used degrees Celsius, but that started at absolute zero. Counting upward from absolute zero, we reach a temperature of 273.15 degrees above absolute zero; that is, 273.15 degrees kelvin. Kelvin's name is not capitalized in the SI system, but the symbol is K, so that one can say 273.15 K. The degree symbol (°) is not used in this case.

In the SI version, the Kelvin scale is official, for that is the most convenient for scientists, who find that using it simplifies physical and chemical relationships. The Celsius scale, however, is permitted for ordinary use as well.

We start Step 1 at 1 degree Celsius rather than at 0 degrees Celsius because we can't very well multiply or divide zero by 3.16 and get any answer other than zero.

A temperature of 1 degree Celsius (or 33.8 degrees Fahrenheit) is quite familiar to anyone in the temperate zones of the world. The atmosphere is generally at that temperature some time during the course of an ordinary winter day in New York City, for instance.

STEP 2

3.16 degrees Celsius ($10^{0.5}$ °C)
276.31 kelvin ($10^{2.441}$ K)

Ice is less dense than liquid water. Since ice is made up of the same molecules that water is, this must mean that the molecules in ice are more loosely organized than in water, more widely spread apart, on the average.

When ice melts, the molecules fall into a more compact arrangement and the density goes up. However, there is still a little "iciness" to the arrangement, so to speak. As the temperature continues to go up, the iciness diminishes, and the density continues to go up slightly.

A rising temperature, however, gives the molecules of water more energy on the average, so that they move more quickly and push each other farther apart. That tends to lower the density. Eventually the tendency of the rising temperature to lower the density surpasses the effect of disappearing iciness to raise the density.

It is at a temperature of 3.98 degrees Celsius—just about the Step-2 temperature-level—that the former overtakes the latter, and water reaches its maximum density under Earth-surface conditions.

This is a fact of great importance for life. In those portions of the Earth where winter conditions lower the temperature toward the freezing point of water, the surface waters in rivers, ponds, lakes, and the ocean itself drop in temperature and grow denser than the water beneath. The surface waters sink and are replaced by the warmer and less dense water below which, in its turn, is cooled and made to sink.

Finally, all the water reaches a temperature of 3.98 degrees Celsius. When the water at the surface cools below that temperature, it actually decreases slightly in density, and does not sink. It is only the water on the surface that freezes, therefore, and the ice, being less dense still, floats on top of the water and finally forms a solid topping, insulating the water below. It is for this reason that sizable bodies of water do not freeze solid in even unusually cold winters, and this allows life in such bodies to continue.

STEP 3

10 degrees Celsius (10^{1} °C)
283.15 kelvin ($10^{2.452}$ K)

The usual metric prefixes are not used in temperature measurements so that one does not speak of a "dekadegree" or a "kilodegree," or, for that matter, a "millidegree." One simply speaks, at the Step-3 temperature-level for instance, of 10 degrees Celsius.

This is equivalent to 50 degrees Fahrenheit, which is the temperature reached on an early spring day in New York City, a temperature that feels pleasant enough after the cold of winter, though it would feel chilly if it came in the early fall after the heat of summer.

The average temperature of the Earth's surface, day and night, winter and summer, polar regions and tropics, is about 14 degrees Celsius (or 57 degrees Fahrenheit), though there are few places that aren't considerably higher, or lower, than that at most times.

STEP 4

31.6 degrees Celsius ($10^{1.5}$ °C)
304.75 kelvin ($10^{2.484}$ K)

The Step-4 temperature-level is equivalent to just about 89 degrees Fahrenheit, and this would be encountered on a hot summer day in New York City.

The normal temperature of the human body is 37 degrees Celsius (or 98.6 degrees Fahrenheit). The chemical reactions always taking place within the human body continually produce heat, so that body temperature is maintained even though heat is continually being lost to the outside world—which is usually at a lower temperature.

Cold-blooded animals (that is, all animals but birds and mammals) usually have their body temperature equal to the environmental temperature about them. This means that, as the temperature drops at night or in the winter, the energy content of the body sinks and the cold-blooded land animal grows sluggish. (Water animals that are cold-blooded, such as fish, are adapted to a surrounding temperature that is *always* cold and are lively enough.)

Sluggishness is a disadvantage, but, on the other hand, cold-blooded animals can get by on a comparatively small food supply since they don't have to burn food continuously just to keep the body temperature high. Warm-blooded animals (such as the human being) can be active at all times, but have to have a generous food supply if they are to keep themselves warm.

If the temperature is very low, the loss of body heat to the cold surroundings proceeds at such a rapid rate that no amount of food will do the job, and the animal will freeze. The rate of heat loss is reduced, however, by an insulating layer, such as feathers in birds, hair in mammals, or artificial layers of clothing in human beings. (Human beings can also raise the temperature of the environment by the use of fire.)

If the temperature is unusually high, on the other hand, it becomes difficult for the heat, constantly generated within the body, to be radiated away. (It is for that reason that a temperature of 24 degrees Celsius probably feels more comfortable than temperatures either below or above.)

Human beings don't rely on the simple loss of heat by radiation and similar processes. They also perspire, so that a film of water is delivered to the surface of the body. The water evaporates, and this process absorbs heat (for it takes energy to pull the molecules of liquid water apart, and to form vapor out of them). The heat needed for the purpose is withdrawn from the body surface, which thus cools off. Perspiration is the body's air-conditioning system.

If there is considerable vapor in the air already, further evaporation is slowed. Perspiration is produced faster than it evaporates, liquid collects on the skin, and we sweat visibly—the air-conditioning system then fails to work properly. That is why a temperature of 30 degrees Celsius or so can be borne if the air is quite dry, but becomes difficult to support if the air is humid.

Earth's temperatures on the planetary surface rarely go much above the Step-4 level. The record high temperature in New York City was 41.1 degrees Celsius (106 degrees Fahrenheit) in the shade. On Earth generally, a temperature of 58 degrees Celsius (136 degrees Fahrenheit) in the shade was once recorded in Libya.

STEP 5

100 degrees Celsius (10^{2} °C)
373.15 kelvin ($10^{2.57}$ K)

At 100 degrees Celsius, we are at the boiling point of water. Such a temperature does not occur on the surface of the Earth, except where material is brought upward from lower depths, as in hot springs and volcanoes. At this temperature-level, we can no longer deal with life as we know it (though some bacteria that live in hot springs can tolerate temperatures near the boiling point).

The Moon is, on the average, as distant from the Sun as Earth is, but the Moon has no atmosphere to absorb the sun's heat and, by means of air currents, to distribute it around its sphere. Then, too, whereas any place on Earth's surface remains in sunlight for only about 12 hours at a time, on the average, places on the surface of the much more slowly rotating Moon can remain in sunlight for more than 350 hours at a time. As a result, places on the Moon's surface can, at the Lunar noon, reach temperatures of up to 117 degrees Celsius, well above the boiling point of water. (To be sure, there is no water on the Moon to undergo boiling.)

STEP 6

316 degrees Celsius ($10^{2.5}$ °C)
589.15 kelvin ($10^{2.77}$ K)

There are hot springs on the ocean floor that can reach 350 degrees Celsius under the huge pressures that exist there.

If we move up into the atmosphere, the temperature drops below the surface values, at first. However, the higher we go, the less dense the atmosphere, and eventually, what sunlight is absorbed, being distributed among far fewer particles, gives each one a higher energy content than is possible at Earth's surface. The temperature goes up.

At a height of 115 kilometres, the temperature of the atmosphere is at the Step-6 temperature-level. This does not mean, however, that astronauts passing through the region are in any danger. The *total* heat remains small.

If we imagine ourselves burrowing deep into the Earth, the temperature goes up, and the density also goes up. As the temperature rises, the total heat present in a given volume of matter goes up even faster. About 3 or 4 kilometres down into the crust, the temperature will reach Step-6 values, and down there (if we were to send drilling devices that low, for instance) the temperature could not be ignored.

The planet Mercury has, like the Moon, no atmosphere and a very slow rotation. It is, moreover, much closer to the Sun, and its surface temperature can reach as high as 430 degrees Celsius, when the Sun is at its closest and is directly overhead.

The hottest planetary surface in the Solar system is that of Venus. Though farther from the Sun than Mercury is, Venus has an atmosphere that is very dense and that is mostly carbon dioxide, which traps the solar heat and does not allow it to escape readily. The temperature, therefore, builds up beyond anything that exists on the surface of the Moon or of Mercury.

Furthermore, because there is no atmosphere on the Moon and Mercury to distribute heat into the night portion, the temperature on those bodies drops, during the long night periods, to quite low values. On Venus, the thick atmosphere distributes heat efficiently so that the surface temperature is about the same (and intolerably hot) everywhere, from the poles to the equator, both day and night. Everywhere and at all times, the temperature is about 457 degrees Celsius.

At the Step-6 temperature-level, familiar substances change their state.

Mercury, a liquid under ordinary conditions, boils at 356.58 degrees Celsius. Tin melts at 231.89 degrees Celsius, and lead melts at 327.4 degrees Celsius. When the Mercurian surface is at its hottest, mercury, if it existed there in elementary form, would be a vapor, while tin and lead would be liquids. —And this would be true anywhere on Venus's surface at any time.

We have also reached temperatures capable of igniting some common gases. Acetylene (C_2H_2) will catch fire at 335 degrees Celsius.

STEP 7

1 000 degrees Celsius (10^3 °C)
1 273.15 kelvin ($10^{3.10}$ K)

At a height of 200 kilometres above Earth's surface, the faint traces of air present are at a temperature of about 1 000 degrees Celsius, and the same is true of the dense rock 50 kilometres below Earth's surface.

As temperature goes up, the electromagnetic radiation emitted by all bodies above absolute zero becomes shorter in wavelength (on the average). By the time a temperature of 600 degrees Celsius is reached, enough radiation in the longest-wave visible-light region is emitted to make an object appear dully red-hot. It would be brightly red-hot at 1 000 degrees Celsius. As more and more shorter-wave light is added at still higher temperatures, objects would glow orange, then yellow-white, then blue-white.

The air in the upper atmosphere emits too little total radiation to seem to be visibly glowing, but the Earth's substance underground would be red-hot if it were visible—as it is during volcanic eruptions.

Temperature goes up steadily as one penetrates deeper and deeper into the interior of any sizable astronomical body. On the whole, the temperature goes up more slowly the smaller the body. Even on the Moon, though, which is considerably smaller than the Earth, the central temperature may be at about 1 600 degrees Celsius.

Such temperatures need not always be safely hidden within a planetary interior. The Sun is slowly growing hotter as it evolves. When it finally swells up to a large "red-giant" star, some seven billion years hence, the Earth's surface (and the Moon's too) may attain a temperature as high as 1 000 degrees Celsius. Earth would have become uninhabitable at least a billion years earlier.

Silver melts at 960 degrees Celsius, gold at 1 063 degrees Celsius, and copper at 1 083 degrees Celsius.

Human beings learned how to produce and use temperatures at the Step-7 level with the discovery of fire. Fires using wood, coal, or oil as fuel yield temperatures of from 1 500 to 2 000 degrees Celsius.

STEP 8

3 160 degrees Celsius ($10^{3.5}$ °C)
3 433.15 kelvin ($10^{3.536}$ K)

At a depth of 1 megametre below the Earth's surface, the temperature reaches the Step-8 level. By the time the very center of the Earth is reached, the temperature is at least 4 000 degrees Celsius and is possibly as high as 6 000 degrees Celsius.

This is higher than the temperature at the surface of some stars. A dim red dwarf, such as Proxima Centauri, may have a surface temperature of as little as 2 400 degrees Celsius. The temperature of the surface of the Sun is about 5 500 degrees Celsius.

The sunspots are cooler than the rest of the Solar surface (which is why they look black in contrast to the hotter, unspotted surface, though they are not truly black, but would gleam brightly if they could be seen in isolation). Sunspots may have temperatures as low as 4 000 degrees Celsius at their center.

On the whole, then, the Earth's core is as hot as the Sun's surface.

Human beings can produce flames with a Step-8 level of temperature. Acetylene, burning in air, will yield a flame with a temperature of about 2 400 degrees Celsius, just about that of the surface of a small star. In pure oxygen, an acetylene flame will have a temperature of 3 100 degrees Celsius. Cyanogen (C_2N_2), burning in oxygen, will yield a flame of 4 500 degrees Celsius, while carbon subnitride (C_4N_2) will yield one of 5 250 degrees Celsius. This is the hottest flame produced by human beings through chemical combustion, and it is very nearly the temperature of the Sun's surface.

There are very few solids or liquids that can remain in such form at Step-8 levels of temperature. The metal osmium melts at 2 727 degrees Celsius and boils at about 4 100 degrees Celsius. A lump of osmium would manage to remain solid near the surface of a small red dwarf star, but would vaporize near the surface of the Sun.

The metal with the highest melting point is tungsten, which melts at about 3 415 degrees Celsius and boils at about 5 000 degrees Celsius. An oxyacetylene torch would not quite melt tungsten, but even that metal would be a vapor at the surface of the Sun. There are a few compounds, such as tungsten carbide (WC), that don't boil till a temperature of about 6 000 degrees Celsius is reached, but beyond that point, all known substances are in gaseous form—at low pressure, that is.

The interiors of the Earth and other planetary bodies, even when they are at temperatures of 6 000 degrees Celsius and above, are, nevertheless, liquid, or even solid, because the matter there exists under enormous pressures.

STEP 9

10 000 kelvin (10^4 K)

At this point, it is no longer useful to distinguish between the Celsius scale and the kelvin scale. A difference of 273.15 degrees out of 10 000 is less than 3 percent, and it grows still less as the temperature rises higher. At the Step-9 temperature-level and above, we might as well choose one of the scales, and since the kelvin scale is preferable in the scientific view, we will cling to that.

Main-sequence stars that are more massive than the Sun are hotter on the surface. The star Sirius, which is nearly three times as massive as the Sun, has a surface temperature of about 10 000 kelvin.

White dwarf stars are far smaller than the Sun in bulk, and may be no more massive, but they too, if they are young enough, can have a hotter surface. Sirius B, the white-dwarf companion of Sirius, has a temperature of 10 000 kelvin, just as Sirius does.

On the Sun, just as there are places where the surface temperature is lower than normal, there are places where it is higher than normal. Occasionally, there are energetic explosions on the Solar surface that result in "flares." These flares, which usually endure only from a few minutes to an hour, have temperatures of 15 000 kelvin or so.

STEP 10

31 600 kelvin ($10^{4.5}$ K)

The largest and most massive stars to be found on the main sequence have surface temperatures in the Step-10 level and only rarely reach the mark of 40 000 kelvin. A white dwarf star, freshly formed, may have a surface temperature of 50 000 kelvin.

The surface of a star is its coolest portion, however. The temperature rises as one moves down into its interior, and also as one moves up into its atmosphere (as is true on Earth). At a height of some 1.5 megametres above the Sun's visible surface, for instance, the temperature is about 20 000 kelvin. Of course, the total heat of a given volume of Solar atmosphere goes down with height, for though the individual particles contain more energy, the total number of particles in a given volume declines more sharply than the energy content of each particle rises.

The core of a small planet, such as Earth, can rival the surface temperature of an average star such as the Sun, but we must expect a large planet to do better. The central core of Jupiter may have a temperature as high as 54 000 kelvin, rather hotter than the surface of any main-sequence star.

If we continue in single steps, we can imagine ourselves higher in the atmosphere and deeper in the interior of the Sun and other stars, but this would scarcely be exciting. Let us skip a bit.

. . .

STEP 13

1 000 000 kelvin (10^6 K)

The corona is the Sun's outermost atmosphere, and is visible (without special equipment) only during a total eclipse of the Sun. It then shines with a total radiance about that of the full Moon. Its dim light could easily make one imagine that it is a relatively cool vapor, but it is not. The dimness is a function of the low density of the gas, only a hundred-trillionth, at most, of Earth's sea-level atmospheric density. Its temperature is 1 000 000 kelvin at its lowest point, and is higher still as one recedes from the Sun.

STEP 14

3 160 000 kelvin ($10^{6.5}$ K)

There are places in the corona where the temperature rises to about 4 000 000 kelvin. There are "hot spots" in space near Jupiter and Saturn, where thinly spread particles, powered by the huge magnetic fields of those giant planets, attain even higher temperatures, if we judge by the energies of those very occasional particles. Space vehicles pass through such hot spots without being adversely affected.

A much more important example of such temperatures, with densities and pressures very high, is in the deep interiors of stars. At a depth of 600 megametres below the Sun's surface, for instance, the temperature is estimated to be 6 000 000 kelvin.

STEP 15

10 000 000 kelvin (10^7 K)

We are now in the temperature range of the centers of stars. The central temperature of a small red-dwarf star would be as high as 8 000 000 kelvin, while that of our Sun is probably at 15 000 000 kelvin.

A neutron star is very much like a massive stellar core that exists without a surrounding shell of relatively normal matter. Its surface temperature is estimated to be about 8 000 000 kelvin, rather like that at the core of ordinary stars.

STEP 16

31 600 000 kelvin ($10^{7.5}$ K)

Perhaps surprisingly, scientists have managed to achieve temperatures higher than those at the center of the Sun. That is, they have treated thin wisps of hydrogen with electromagnetic fields, with laser beams, and so on, in such a way as to give the very few hydrogen nuclei present enormous individual energies. This they have been doing in search of controlled nuclear fusion, in order to make available to human beings the same sort of energy the Sun gives us, but in places and quantities of our own choosing.

The highest temperatures so far achieved are about 67 000 000 kelvin, well past the Step-16 level, and over four times that at the core of the Sun. (The core of the Sun has enormous pressures working toward fusion; scientists don't. Scientists must therefore attain temperature levels much higher than those at the Sun's center, to make up for the fall-short in pressure, if controlled fusion is to be developed.)

STEP 17

100 000 000 kelvin (10^{8} K)

If we imagine ourselves looking backward in time, we would find ourselves dealing with a smaller and smaller Universe, and approaching closer and closer to the Big Bang, when the Universe had all its mass crammed into an infinitesimal volume. As we moved farther and farther back in time, we would be dealing with a Universe in which the total energy was squeezed into a smaller and smaller volume, so that there would be a higher and higher temperature. If we look back far enough, we can expect to deal with temperatures higher than those that occur even in stellar interiors.

Thus, some 3 hours after the Big Bang, the Universe was so small that all of it might have fitted inside the orbit of the Earth, and its average temperature would have been 100 000 000 kelvin.

STEP 18

316 000 000 kelvin ($10^{8.5}$ K)

Human efforts at producing temperatures for controlled fusion are outdone by the temperatures we have created in setting off uncontrolled fusion. At the center of a large hydrogen bomb, it is estimated that temperatures up to 400 000 000 kelvin are reached.

STEP 19

1 000 000 000 kelvin (10^9 K)

Even a hydrogen bomb is not the ultimate, of course. The core of a neutron star may have a temperature as high as 800 000 000 kelvin.

. . .

STEP 21

10 000 000 000 kelvin (10^{10} K)

Exploding stars are, in a way, exploding hydrogen bombs, but are enormously vaster and more energetic than anything human beings can conceivably bring about on Earth's surface. A large supernova can attain temperatures (briefly) of as much as 10 000 000 000 kelvin.

One second after the Big Bang, when the Universe may have been smaller in volume than our Sun is today, its average temperature is thought to have been 10 000 000 000 kelvin.

. . .

STEP 25

1 000 000 000 000 kelvin (10^{12} K)

A ten-thousandth of a second after the Big Bang, when the Universe was no larger in volume than a large asteroid is today, its average temperature is thought to have been 1 000 000 000 000 kelvin.

. . .

STEP 65

100 000 000 000 000 000 000 000 000 000 000 kelvin (10^{32} K)

Present physical theories do not allow physicists to estimate conditions at less than 10^{-43} seconds after the Big Bang. At that time, the entire Universe was far, far less voluminous than a single proton of the kind we know today, and its temperature is thought to have been at the Step-65 level of 100 000 000 000 000 000 000 000 000 000 000 (a hundred million trillion trillion) kelvin. With that, we are forced to stop the climb up the ladder of temperature.

We have gone from the temperature of a winter day to that of the near neighborhood of the Big Bang in 64 steps, covering 32 orders of magnitude.

The Ladder of Temperature

DOWNWARD

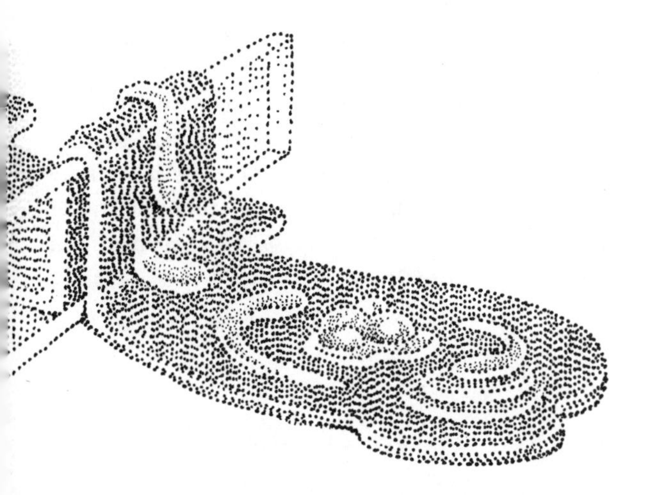

DOWNWARD

STEP 1

316 kelvin ($10^{2.5}$ K)

In moving down the ladder of temperature, we might go back to 1 degree Celsius, which is where we started the upward climb. In multiplying that figure by 0.316 at each step as we have done in earlier descending ladders, we would merely get closer and closer to 0 degrees Celsius, and would never pass below it. The same would be true if we start at *any* temperature above 0 degrees Celsius.

We might therefore start at −1 degree Celsius and try to multiply *that* by 0.316. Again we would move closer and closer to 0 degrees Celsius, and that would be true for *any* negative temperature.

For that reason, let us switch to measurements in the kelvin scale as we did for very high temperatures. We will start at 316 kelvin.

A temperature of 316 kelvin (42.85 degrees Celsius, or 109.13 degrees Fahrenheit) is close to the maximum temperature found in the shade on Earth's surface. We have already considered this region on the upward climb and we won't linger; we will descend.

STEP 2

100 kelvin (10^2 K)

A temperature of 100 kelvin is equal to −173.15 degrees Celsius, so that in moving from Step 1 to Step 2, we pass through the entire temperature range at or near the Earth's surface.

The coldest temperature New York City has experienced in Weather Bureau history was 247 kelvin (−25 degrees Celsius, or −14 degrees Fahrenheit).

The coldest temperature anywhere on the Earth's surface outside Antarctica is at Verkhoyansk, Siberia, where a temperature of 203 kelvin (−70 degrees Celsius, or −94 degrees Fahrenheit) was recorded. At Vostok, the Soviet base in Antarctica, which is the coldest part of that frigid continent, a temperature of 185 kelvin (−88 degrees Celsius, or −127 degrees Fahrenheit) was observed.

Although the temperature of the Earth's atmosphere is very high in its upper regions, it is low in the lower regions, and the low point is at a height of about 85 kilometres. At that height, a temperature of 130 kelvin (−143 Celsius, or −225 Fahrenheit) was at one time observed. This is the coldest temperature ever observed on Earth, and it approaches the Step-2 temperature-level.

Other planets in the inner Solar system don't do much better. Mercury, which is so hot on the surface lying under its swollen Sun, has a nighttime period that is about an Earth-month long. Its surface loses temperature to outer space, and by the time the long-delayed morning arises and the Sun is about to broil the surface once more, a low temperature of about 150 kelvin has been reached (considerably colder than the coldest Antarctic night).

Mars is farther from the Sun than Mercury is, but its night is only a little over 12 hours long, and by morning its surface is at a temperature of 170 kelvin. The Moon, with a two-week night, attains a temperature of 110 kelvin by morning.

The outer planets experience low temperatures even when the Sun is shining. The temperature of Jupiter at the visible cloud-layer surface is about 120 kelvin, for instance. To search for temperatures lower than the Step-2 level, then, we must reach out beyond Jupiter.

At low temperatures, substances we think of as liquids freeze to solids, while substances we think of as gases, condense to liquids, and eventually solidify as well.

Among the liquids, for instance, mercury freezes at 234.3 kelvin and ethyl alcohol at 158.6 kelvin.

Among gases, chlorine liquefies at 239.1 kelvin, and solidifies at 172.1. Xenon, one of the noble gases found in small quantities in the atmosphere, liquefies at 165 kelvin and solidifies at 161.3 kelvin, while butane (a gas commonly used as fuel) liquefies at 261.5 kelvin and solidifies at 113.6 kelvin.

At 100 kelvin, eight of the substances we think of as gases remain as gases: argon, fluorine, helium, hydrogen, neon, nitrogen, oxygen, and carbon monoxide (CO). Such gases will all remain gases on all those worlds of the Solar system that never move outward farther than the orbit of Jupiter.

STEP 3

31.6 kelvin ($10^{1.5}$ K)

At this temperature-level, we must consider the outer Solar system. The visible cloud-surface of Saturn is at about 90 kelvin; that of Uranus, at 65 kelvin; that of Neptune, at 50 kelvin; and that of Pluto, at its farthest from the Sun, at perhaps 40 kelvin. None of the planetary bodies, therefore, attain a Step-3 temperature-level.

There are, however, comets that recede farther than Pluto's orbit or that (it is thought) circle in vast orbits every part of which is far beyond that of Pluto. These may exist at temperatures well below 31.6 kelvin, and since they are small bodies, only a few kilometres across, they would be at such low temperatures through and through.

Some of the gases that are still gases at the Step-2 temperature-level, liquefy, and even solidify at the Step-3 temperature-level. Thus, fluorine liquefies at 85.0 kelvin and solidifies at 83.8 kelvin; oxygen liquefies at 90.2 kelvin and solidifies at 54.4 kelvin: nitrogen liquefies at 77.4 kelvin and solidifies at 63.2 kelvin; and carbon monoxide liquefies at 81.7 kelvin and solidifies at 68.2 kelvin.

All these gases would be solids on Pluto.

There are just three gases that are still gases at 31.6 kelvin; they are hydrogen, helium, and neon. These gases would remain gases on the surface of every known body in the Solar system.

Scientists worked all through the nineteenth century to attain low temperatures, and, by the last decade of that century, finally attained temperatures lower than those on the surface of Pluto (which had not yet been discovered at that time).

Scientists discovered that very cold substances could possess properties that were never observed at ordinary temperatures. One of these is "superconductivity." A substance is superconductive when it offers zero resistance to an electric current. An electric current set up in a ring of superconductive matter will continue traveling through the ring forever, if it is kept below a certain "transition temperature" and if it is left otherwise undisturbed.

The highest temperature at which superconductivity has been observed, so far, occurs in the case of an alloy of niobium, aluminum, and germanium, where the transition temperature is 23 kelvin.

STEP 4

10 kelvin (10^1 K)

By the time we sink to the Step-4 temperature-level, two of the three last gases have given in. Neon liquefies at 27.0 kelvin and solidifies at 24.6 kelvin, while hydrogen liquefies at 20.4 kelvin and solidifies at 14.0 kelvin. At 10 kelvin, only helium remains as a gas.

The highest transition temperature for superconductivity in any pure element is that of niobium, for which it is 9.26 kelvin.

STEP 5

3.16 kelvin ($10^{0.5}$ K)

Now even helium gives way. Helium liquefies at a temperature of 4.2 kelvin. At 3.16 kelvin, it has not yet solidified, however, but remains a liquid. It is the only substance that can, at this temperature-level, exist as anything but solid.

Helium exists in two isotopic varieties, helium-4 and helium-3. Only a very small fraction of helium, as it exists in nature, is helium-3. If helium-3 is isolated and collected, it turns out to liquefy at an even lower temperature than helium-4. The liquefaction point of helium-3 is 3.2 kelvin—almost exactly the Step-5 temperature-level.

The average temperature of the Universe is about 3 kelvin.

Suppose a body were located in space between the galaxies and received nothing more from the Universe than very faint starlight, cosmic ray particles, and so on. If that body contained no energy at all to begin with, it would absorb some of the radiation and particles and would gain energy—and would, therefore, begin to re-radiate some of that energy to space.

Eventually, an equilibrium would be reached. The body would re-radiate energy as quickly as it absorbed it, and the equilibrium temperature at that point would be about 3 kelvin.

This means that it is very unlikely that there is anything in the Universe which is at an equilibrium temperature of less than 3 kelvin. Lower temperatures were produced in human laboratories in the 1920s, and such temperatures are not likely to exist anywhere else in the Universe except, possibly, in the laboratories of intelligent beings on other worlds, if any such exist, with technologies as advanced as ours.

The first case of superconductivity discovered was that of mercury, by the way. Its transition temperature is 4.15 kelvin.

STEP 6

1 kelvin (10^0 K)

At a temperature of 1 kelvin, helium is still liquid. In fact, it doesn't freeze at all (at ordinary pressures), no matter how low the temperature gets. Even at absolute zero, a tiny trace of energy is still left in substances. The temperature is absolute zero, because at that temperature no further energy can be removed by any conceivable method, even in theory, but that last tiny trace of *irremovable* energy is enough to keep helium liquid.

It was found, however, that at 1 kelvin and below, placing a pressure of about 25 atmospheres upon liquid helium would cause it to solidify, so that solid helium *can* exist, even if not under atmospheric pressure.

There are about 17 elements known to become superconductive in the neighborhood of 1 kelvin. Gallium, for instance, has a transition temperature of 1.091 kelvin. Thallium has one that is somewhat higher, 2.39 kelvin; osmium, one that is somewhat lower, 0.655 kelvin.

At a temperature of 2.2 kelvin, helium changes from a liquid of normal properties (helium I) to one of completely unprecedented ones (helium II). As one example of its strange properties, helium II is "superfluid" and can move through very small orifices without any measurable friction. A system might be gas-tight and yet not be helium II–tight.

Helium II was found to involve helium-4 only. That, is helium-3 retained ordinary properties even at temperatures as low as 1 kelvin.

STEP 7

0.316 kelvin ($10^{-0.5}$ K)

Original attempts by scientists to attain low temperatures involved, in essence, making use of evaporation in one way or another. Evaporation, as I mentioned earlier, is a process that absorbs heat, and, if the situation is properly handled, the heat absorbed comes from the body of the liquid that is evaporating.

In the 1920s, this process came to a dead end with temperatures of 0.4 kelvin—nearly at the Step-7 level. This would be enough to attain the transition temperature for ruthenium, which is 0.49 kelvin, but not quite enough to catch that for titanium, which is 0.39 kelvin.

But then, in the 1920s, magnetic methods were used which could lower the temperatures attainable quite a bit. So far, four elements are known that become superconductive only at temperatures below 0.4 kelvin. Perhaps, if elements could be studied at temperatures low enough, every one of them—indeed every substance, element or not—might be found to become superconductive.

STEP 8

0.1 kelvin (10^{-1} K)

By the early 1930s, temperatures as low as 0.13 kelvin could be obtained. The transition temperature of iridium is 0.14 kelvin, and that of hafnium is 0.09 kelvin.

. . .

STEP 10

0.01 kelvin (10^{-2} K)

Only one element has been found to have a transition temperature lower than that of hafnium. That is tungsten, which becomes superconductive only at a temperature of 0.012 kelvin.

STEP 11

0.003 16 kelvin ($10^{-2.5}$ K)

The lowest temperature attained by the use of magnetic methods was 0.003 kelvin. To get lower still, scientists used rather subtle methods involving mixtures of helium-3 and helium-4. In that way, scientists found that helium-3 which, earlier, was not thought to take on the helium II properties that helium-4 did, *did* do so, but only at temperatures under 0.002 5 kelvin.

. . .

STEP 15

0.000 031 6 kelvin ($10^{-4.5}$ K)

Making use of the helium-3/helium-4 mixture method, scientists finally reached temperatures as low as 0.000 02 kelvin, and sustained them for a significant period of time.

. . .

STEP 21

0.000 000 031 6 kelvin ($10^{-7.5}$ K)

The lowest temperature scientists have reached at least for a moment has been reported to be 0.000 000 05 kelvin.

Although temperatures only a twenty-millionth of a degree removed from absolute zero have thus been reached, it is useless to expect ever to reach absolute zero itself. Each step toward that goal is as hard to carry through as the one before. Even though less and less heat remains to be extracted, the amount that can be extracted by any conceivable process is always less than all the heat contained.

EPILOGUE

So, by careful and regular stages, a better understanding of the measure of the Universe in a number of its most important physical aspects has been (I hope) brought about.

More astonishing than the vastness of understanding that human thought has achieved is the sheer unlikelihood of this accomplishment and our all-but-unbelievable good fortune in sharing it.

Consider that on Earth it is not likely that consciousness can exist without life, and that among the perhaps twenty million species that have existed in the last three billion years or so, only one, *Homo sapiens,* has experienced consciousness of a kind that could be applied to more than the immediate desires and fears of the moment.

Only the human brain has had the ability to take care of all the material needs of subsistence and yet, thereafter, to have the leftover capacity to concern itself with the Universe in all its abstraction, for no other purpose than to satisfy curiosity and wonder.

Even those few brains that are larger than human—those possessed by elephants, dolphins, and whales—show no signs of concerning themselves (or of being able to concern themselves) with more than their immediate surroundings and needs. We have no evidence of any capacity for forethought and abstraction among them in more than trivial ways.

The same is true for those species most closely related to ourselves, either alive (gorillas and chimpanzees) or dead (the hominids that preceded us, up to and including *Homo erectus*).

We are members of a unique species on Earth, therefore, and since we have no evidence, as yet, for the presence of life elsewhere than on Earth, it is even conceivable that we are the only conglomeration of matter in existence that is capable of considering, measuring, and comprehending the Universe and itself.

Our good fortune does not end there. *Homo sapiens* may have existed on Earth for 350 000 years, but it is only in the last 300 years (less than 1/10 of 1 percent of the whole) that we have had even a vague idea of what the scale of the measure of the Universe might be, and only in the last 60 years or so, that we know the Universe in its apparently* true vastness.

It was only in the 1670s that the distance of the planets came to be known; only in the 1830s that the vaster distance of the stars was understood; only in the 1920s that the still vaster distance of the galaxies was brought within our view.

It was in the 1670s also that the world of microscopic life was first unfolded, but not till the 1770s that the bacteria were first properly studied, and only in the 1890s that the existence of viruses was established.

It was only in the first decade of the 1800s that the existence of atoms was first put on a firm footing; and only in the first decade of the 1900s, that their size was accurately measured, and that the existence of subatomic particles was grasped.

It was not till our own century that the lifetime of the Solar system and of the Universe was calculated, that the speed of light was recognized as a maximum, that the temperatures at stellar cores were determined, that black holes and quarks were contemplated.

So we live in that very small stretch of time during which humanity has finally attained a proper comprehension of the Universe, and we who are now alive are fortunate far beyond our mere involvement in the human fellowship. We might, after all, have lived earlier.

In fact, even among human beings *now,* the vast majority still lack the time, the means, the opportunity (or even, in many cases, the desire) to acquaint themselves with the Universe they inhabit. As a group, only the physical scientists understand the scale of what materially exists. And among them, it may well be that most do so only when they pause to think of it—which may happen rarely, if at all.

*We have, however, been misled before, and it may be that in time to come, additional vastness and intricacy will unfold and we will come to realize that what we now know, or think we know, is but a tiny part of a still greater whole.

Just as human beings, generally, are far too likely to be engaged in the narrow needs of day-to-day subsistence, scientists are far too likely to be engaged in the narrow focusing of day-to-day research.

It may well be, then, that in all of space and time, only a few people —like the readers of this book—have the chance of seriously contemplating the scale of the measure of the Universe (a vision that is peculiarly a part of our own species and our own generation), forcing awe and wonder to stretch to their utmost.

INDEX